Eng
from
First Principles

Engineering Drawing from First Principles

Using AutoCAD

Dennis Maguire
CEng. MIMechE, Mem ASME, R.Eng. Des, MIED

City and Guilds International Chief Examiner in Engineering Drawing

ARNOLD

A member of the Hodder Headline Group
LONDON • SYDNEY • AUCKLAND
Copublished in North, Central and South America by
John Wiley & Sons Inc., New York • Toronto

First published in Great Britain in 1998 by
Arnold, a member of the Hodder Headline Group,
338 Euston Road, London NW1 3BH
http://www.arnoldpublishers.com

Copublished in North, Central and South America by
John Wiley & Sons Inc., 605 Third Avenue, New York
NY 10158–0012

British Library Cataloguing in Publication Data
A catalogue record for this book is available from the British Library

Library of Congress Cataloging-in-Publication Data
A catalog record for this book is available from the Library of Congress

ISBN 0 340 69198 0
ISBN 0 470 32365 5 (Wiley)

Commissioning Editor: Sian Jones
Production Editor: James Rabson
Production Controller: Rose James
Cover designer: Terry Griffiths

Typeset in Meridien by J&L Composition Ltd, Filey, North Yorkshire
Printed and bound in Great Britain by The Bath Press, Avon

Contents

Preface

Thank you for your interest. My first introduction to the subject of technical drawing came at school, when at the age of 11, we started woodwork lessons. At the first meeting of the class Mr Munday, our teacher, told us that we would draw working plans of a simple joint. We were introduced to a drawing board, tee square, set of instruments, one sheet of snowy white paper and two paper clips. We were shown the principles of simple projection and then proceeded to copy views of a Lap Joint from the blackboard. I instantly found this to be a very agreeable relaxing exercise. The controlled use of the pencil, rule, square and compasses also enabled those of us without natural skills in freehand art and painting to be 'in with a chance'.

At the next lesson we were taught how to prepare the wood, cut and make the two parts and finally check the accuracy of the assembled joint. In addition we could see every edge and corner represented by lines on the drawing.

This particular subject at school was especially popular with the other lads and the general class atmosphere very stimulating and rewarding. When the results compared favourably with the initial drawings I felt that I had really achieved something and soon developed a strong desire to continue studies. We all also took great pride in trying to keep the paper nice and clean and free of pencil smudges. I also always enjoyed the subject of Applied Geometry which is logical and required more constructional work on the drawing board.

When I left home in the mornings for school it was not uncommon to see a new neighbour of ours leaving for work. He was smartly dressed and carried a leather briefcase. Out of curiosity I asked my mother what Mr Oram did for a living and she found out that he was a draughtsman and made 'plans'. I decided that this was quite a likely career for me and it certainly gave me some sense of purpose at school when I realised that you could actually get paid good money for doing something you really enjoyed. I later bought a small drawing board and tee square for use at home and copied examples from books in our local library and found that regular practice soon improves technique.

Pencil drawings are easily corrected but damage to the paper may arise. More important drawings for manufacturing and publication purposes were required to be produced in ink. This was a skill requiring lots of patience because ink is reasonably permanent and extensive errors on some surfaces needed a redraw. In industry, it was not uncommon for ink tracers to be employed to trace over pencil drawings onto transparent film after original designs had been produced, checked and approved. These days with the aid of a computer it is not necessary to worry about the state of the paper or the need to trace the finished designs to obtain a truly professional result. No need to keep sharpening the pencil to obtain consistent line thicknesses, no need to worry about spelling mistakes after you have inked in notes when you have otherwise nearly finished your work of art.

Variations in the height and slope of letters and numbers during dimensioning always used to separate the beginner from the professional and often spoil an otherwise sound drawing. These days lines and letters are all standardised and a spell checker available at the touch of a button, but it is still necessary for the draughtsman to be completely in charge. We have not reached the stage where the computer can dispense with the draughtsman. This fact needs to be broadcast very widely and regularly. It is definitely not sufficient to only become a proficient keyboard operator. Computers cannot conceive and produce original designs and drawings. With the aid of a computer though you can produce clear unambiguous drawings provided you adhere to the appropriate National and International Standards. This is the area where I hope to assist you and I am convinced that to develop your talents you must keep your hands on the keyboard.

In the following pages are exercises which I am sure will put some basic knowledge and skills into your fingers and also help generate some draughting speed. Do not worry or get disheartened if the results are a little slow in coming – with computing it happens to us all. As the title suggests, this book is about the production of drawings from basics. All of these drawings have been prepared with the aid of the computer but of course there is no reason why you could not redraw them on a drawing board if you so wish, as this would certainly develop manual ability and gain a clear understanding of the principles of Engineering Drawing. Not all drawing offices use CAD.

I started to teach this subject at evening classes in Hayes and the first drawing office I walked into had the following words printed on a strip of paper above the blackboard: 'Little things make perfection, but perfection is no small thing'. As far as draughtsmanship is concerned there is a lot of truth in this short statement. In themselves, the rules are relatively simple but they need to be employed accurately and consistently, and applied thoroughly. The draughtsman also needs to be a quick and efficient operator and hence the value of practice.

The computer provides a range of facilities which are available for selection by the operator and often allows a job to be done in several different ways. With experience you will make your own pet choice of methods and hopefully achieve the same results. These graded examples start from a completely blank screen. I hope you will soon develop confidence and enjoy draughting with the aid of the computer.

Please note that job vacancies are equally available in drawing offices for both sexes so I wish to apologise for using the widely used collective noun of 'Draughtsman' to cover Drawing Office Personnel. It definitely implies equality of status and it is a pleasure to find an increasing number of young ladies enjoying employment in the drawing office. Thank you again for reading this book. I wish you well and every success in your studies and career.

The British Standards Institution was the world's first national standards body and there are over 80 similar organisations world-wide which belong to the International Organisation for Standardisation (ISO). BSI represents the interests and views of British Industry.

The first edition of BS 308 to cover Engineering Drawing Practice was published in 1927. It is currently published in three separate sections and available in reference libraries throughout the country. The BSI catalogue lists over 10,000 publications and a Yearbook is usually available in reference libraries. Each year new or revised standards are issued to keep the technical contents up-to-date. Revisions encompass new materials, processes and technologies.

Further information on BSI services can be obtained from:

BSI Enquiry Department,
Linford Wood,
Milton Keynes MK14 6LE.
Tel: 0908 221166.

A completely revised edition of *A Manual of Engineering Drawing Practice* has been issued to include an extensive range of important applications of BS and ISO standards which are relevant to current draughting practice. This useful reference book also includes details of general drawing office practice and additional draughting examples from many branches of technology.

Dennis Maguire
January 1998

Acknowledgements

I express my thanks to the British Standards Institution, 389 Chiswick High Road, London W4, for kind permission to reprint extracts from their publications. I have enjoyed a long and enjoyable association of over 35 years with The City and Guilds of London Institute, 1 Giltspur Street, London EC1A 9DD, and am grateful for their kind permission to include past examination questions in this book.

I also thank my friend and colleague Colin Simmons for much helpful information and guidance. The friendly advice and assistance given to me by Dilys Alam, Sian Jones and staff at Arnold is also very much appreciated. My thanks to Ray for much technical help and his invaluable support with my computer equipment, and to Brian for other drawing examples.

Finally many thanks to my wife Beryl for her endless patience, encouragement and all her clerical help.

Chapter 1

First steps

By the end of this chapter, I hope you will be able to understand the following commands and perform these activities:

- Open the drawing software program. Click and browse through each of the options on the menu bar in turn and note that many sophisticated and advanced features are available.
- Reposition the movable toolbox.
- Note standard drawing sizes.
- Appreciate Portrait and Landscape orientation of the drawing.
- Select and position an A4 size drawing sheet.
- Set **Snap** and **Grid** values.
- Draw lines between given coordinates.
- Operate the ZOOM command.
- Become familiar with the ORTHO button facility.
- Experiment with ERASE and BREAK commands.
- Use the REDRAW command to clean up the screen.
- Draw circles and arcs.
- Use the RUNNING OBJECT SNAP OPTION to draw tangents and position tangency points.
- Gain experience with the FILLET feature.
- Convert lines and arcs to polylines for finished outlines.
- Experiment with the TEXT command. Set text types and sizes.
- Note the minimum height of drawing characters and text abbreviations.
- Experiment with the MOVE, COPY and ROTATE commands and learn how to reposition detail.
- Experiment with the ARRAY command.
- Choose screen preferences.
- Select coloured layers for different parts of a drawing.
- Note recommended thicknesses for various types of linework.
- Check available printers. Save your drawings with suitable file names and add drawing numbers.
  You will find that information retrieval is an important part of CAD work. Set up the printer and follow the PRINT/PLOT procedure to make an A4 copy.

Note: In industrial drawing offices the drawings relating to particular contracts are often numbered after an abbreviated contract name followed by the number. Assume that we have a design contract where we would expect 50 or 60 drawings to be needed on a contract from SMITH & CO. Typical drawings could be numbered SMT001, SMT002, SMT003, etc. The sequence of drawings you copy from this book could therefore be ACAD001, ACAD002, etc.

It is also a regular practice to keep a Drawing Register of numbers, titles and completion dates so that other members of staff in an organisation can check on contract progress and the drawings available.

Students often find that starting to learn CAD can be rather slow and the following exercises are therefore designed to introduce gradually interesting aspects of the software applied to useful engineering topics. Both areas of work are progressive so eventually you will possess CAD and valuable engineering draughting experience and be sufficiently skilled to take your place in a modern drawing office. It is also a vital part of the study that you appreciate the reason for the position of every line and possess the ability to edit freely any section of your drawing. A big part of a draughtman's work is to implement modifications and changes to drawings which often result from improvements in design and production techniques.

Initial preparations

I must make the assumption that a software program, for example AutoCAD LT, has been installed on a computer and is ready for your studies, so take a comfortable seat in front of the keyboard, switch on and select the icon on the Windows Program Manager (Figure 1.1).

Double click with the left-hand mouse button over the icon and the screen display shown in Figure 1.2 appears under a central heading of AutoCAD LT - UNNAMED.

Please note that all the following operations are initiated by clicking the left-hand mouse button. Immediately beneath the heading is the pull-down menu line which starts with File, then Edit, Draw, View, Assist, Construct, Modify, Settings and Help. Click over each of them in turn and note the various optional functions that are available to appreciate the scope of the program, but please be assured that you can make rapid progress with only a few commands.

The next line includes the Toolbar which can be personalised to suit your own method of

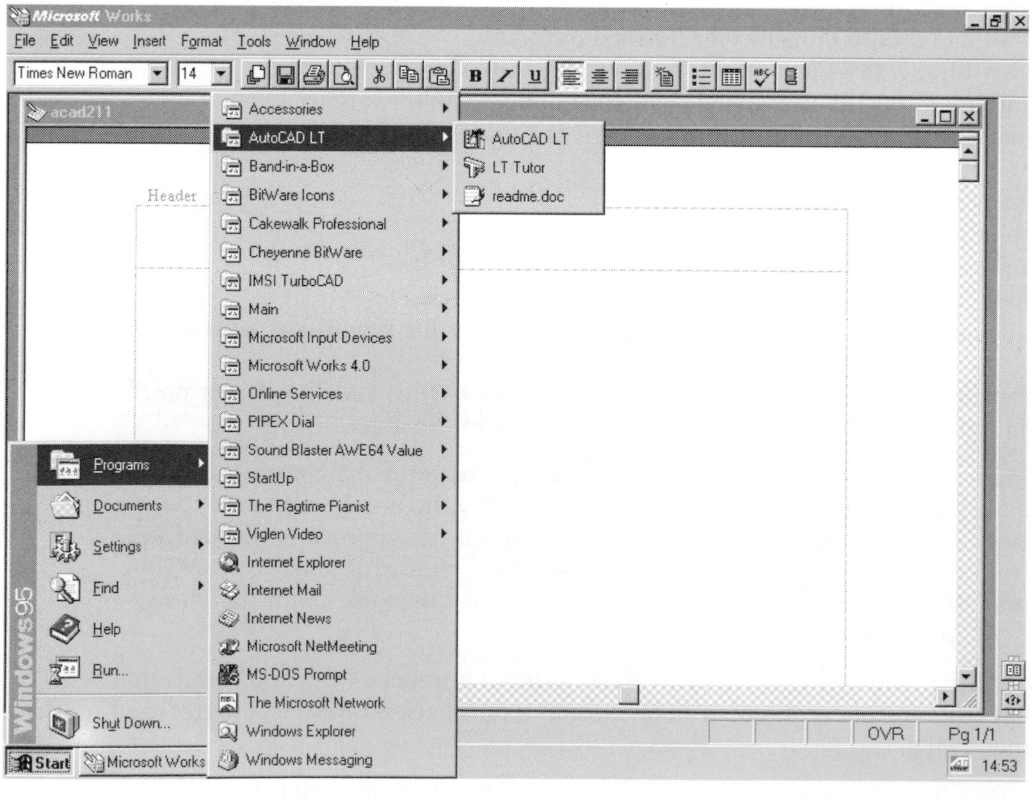

Figure 1.1

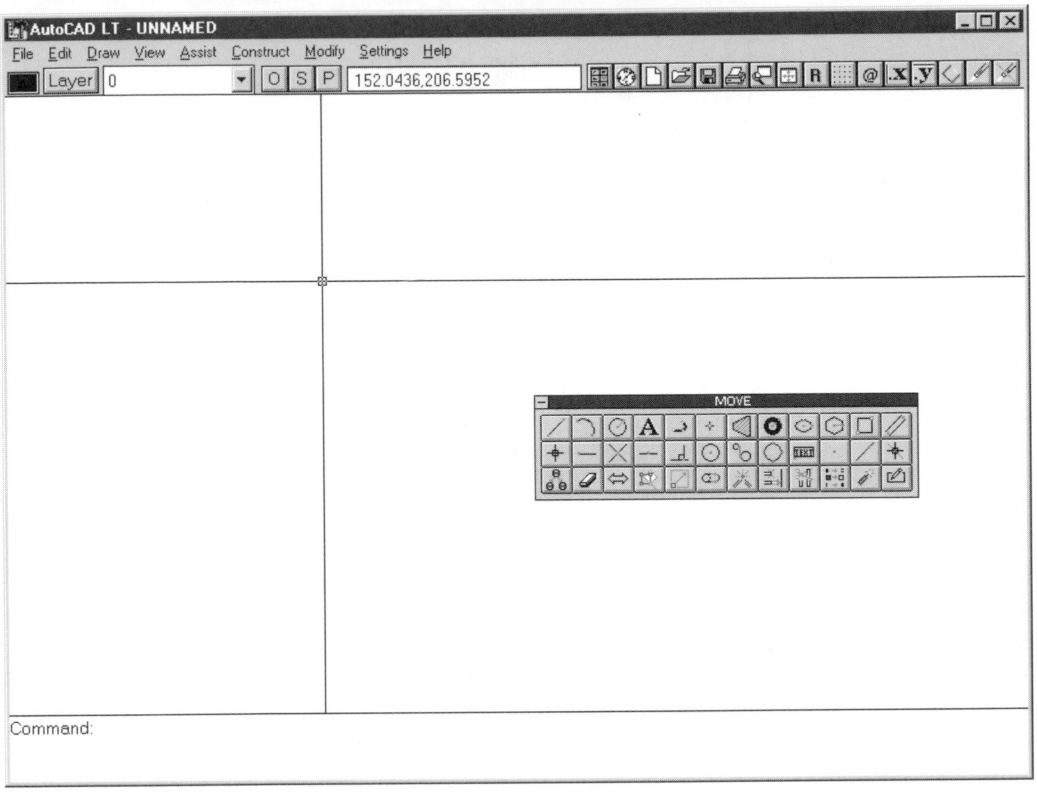

Figure 1.2

working and arranged to hold those commands you most regularly use. In the centre of this line you will find the co-ordinate display where the position of the crosswires on the cursor, which you move with the mouse, are indicated. Two values are given, namely the X and Y co-ordinates.

The graphics or drawing part of the screen area can represent a drawing sheet of any size you wish to use and it is therefore necessary to choose suitable units and scales. It is our intention here to use only metric dimensions.

Setting out lines, arcs and circles

I expect you are keen to produce a meaningful drawing so we need to select a suitable drawing sheet. A4 size is acceptable for most printers with dimensions of 210 mm × 297 mm. You can of course use the sheet in two different ways and 'portrait' style will have the horizontal X dimension of 210 and the vertical Y dimension of 297 mm. Turn the paper round 90° and we have 'landscape' style where the X dimension is 297 and the Y dimension 210 mm. Our computer must first be set for these sizes, so look along the menu list at the top of the screen, click on **Settings** and the menu shown in Fig. 1.3 will appear.

Choose **Limits** and note this message in the command area at the bottom of the screen.

```
Command: _limits
Reset Model space limits:
ON/OFF/ <Lower left corner> <0.0000,0.0000>
```

Press the <Enter> key, which means that you accept this default value, and this next

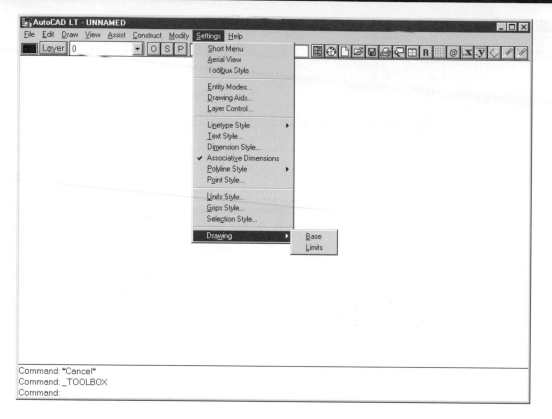

Figure 1.3

message will appear (Note that drawing limits are always quoted with the X and Y values in alphabetical order):

`Upper right corner <420.0000,297.0000>`
This is a default value for a sheet of A3 paper and is double the size we require. Furthermore, it is the wrong way round for portrait style, so type **210,297** and press the <Enter> key.

Note that the area at the bottom of the screen links us with the computer to exchange messages. We cannot indicate our next move until the computer shows `Command:` This message means that it is now ready to receive details of the next operation. After choosing the next event from the dropdown menu, it is necessary to press the <Enter> key and the computer then goes to work. When the operation is complete the `Command:` message will again appear.

Please understand that this symbol <R> or ↵ means that you press the <Enter> key and obviously looks like the key on the keyboard.

Having selected a suitable size of paper on which to draw, we want it to become visible. This is activated via the **Settings** pull-down menu by clicking the **Drawing Aids** dialogue box. The box shown in Fig. 1.4 will appear. Click to turn on the **Grid** which is a series of accurately spaced dots forming a square grid and can conveniently be used for measuring purposes. Adjust the **X Spacing** to 10.00 and you will find that the Y SPACING will adjust itself. Click to turn on the **Snap** feature. This is a most important item, enabling you to position exactly the cursor at specified distances apart and certainly speeds up the accurate placing of points on the screen. I suggest that you set an **X Spacing** of 2.00 and again ignore the Y SPACING. In the **Modes** column at the left-

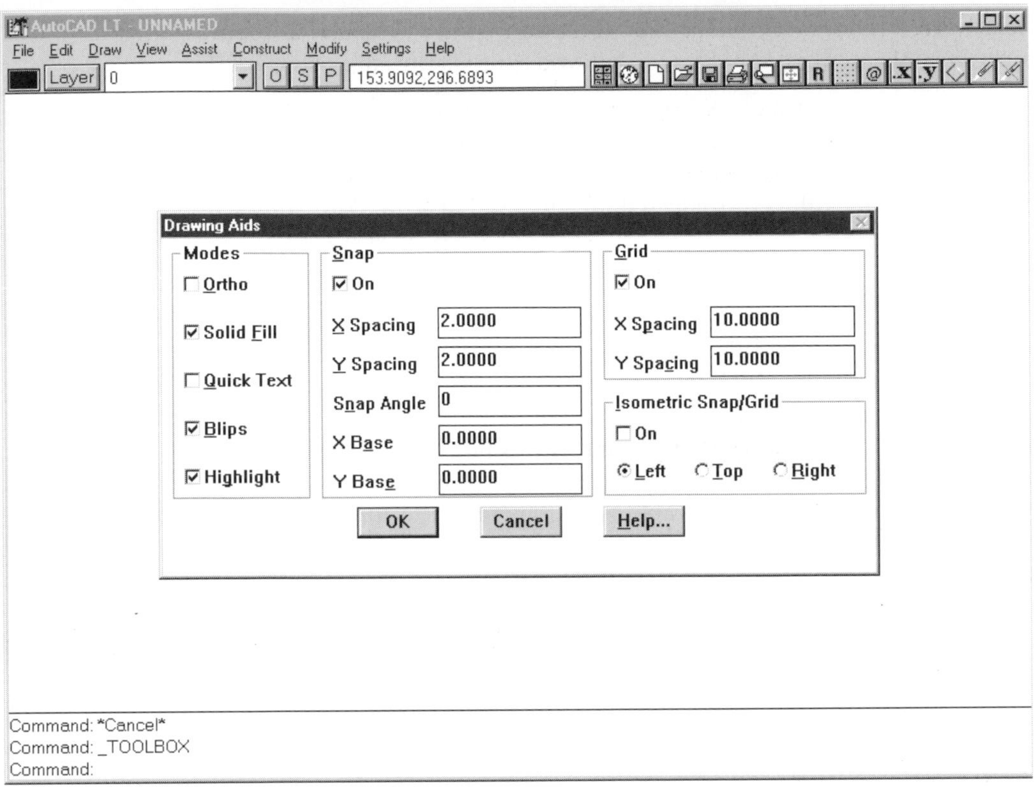

Figure 1.4

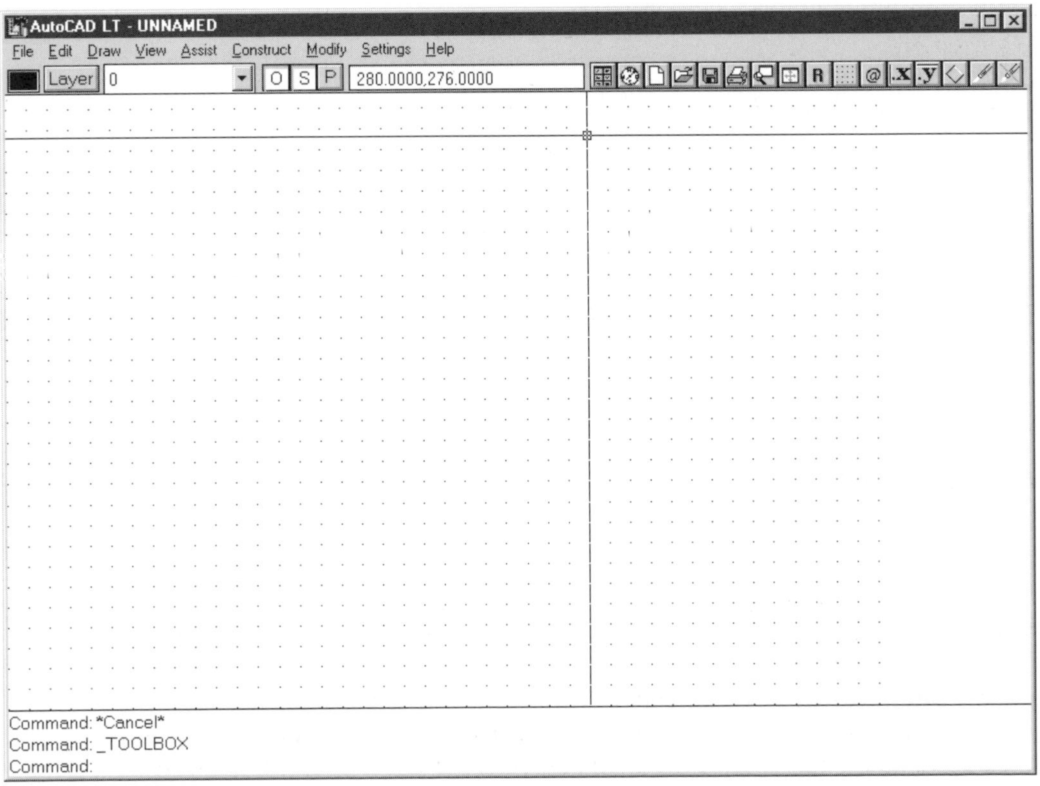

Figure 1.5

hand side click the **Ortho** command. Leave the other values as you find them in their default state. Click **OK**, or press the <Enter> key and a grid of dots now appears (Fig. 1.5).

The screen displays a drawing sheet to represent A4 size and calibrated in 10mm squares. Move the cursor crosswires with the mouse and note the changing values of the co ordinates in the white box above the drawing area. The numbers will change in 2mm increments and this is the value you selected for the SNAP feature. You will now be able to position features accurately on the screen in 2mm steps. If you click on the **Snap** button and turn it off, you will find that the co-ordinate values change and are now given to four places of decimals.

It is not necessary to go through this setting-up procedure each time you start a drawing since a basic sheet can be held in the database. When the drawing is completed you may possibly wish to add a border to give a good professional finish, and notes are given later in the text. A standard border may be conveniently positioned on a separate transparent layer since you do not necessarily want the printer to reproduce the full sheet until the details have been completely edited and approved. Several prints may be required during the preliminary design stages of original projects. Before we start our first drawing, click on the **Zoom** button in the menu bar, shown by the icon of a magnifying glass over a piece of paper. In the command line you will read the following.

```
Command: Zoom
All/Centre/Extents/Previous/Windows/<Scale(X/XP)>:
```
Click at any position on the grid and the command line will now display the instruction `Other corner:`. Click again a short distance away and at an angle. A rectangle is formed and the grid enclosed by that area will be enlarged. Having completed the operation, the computer is ready for the next operation and displays the word `Command:`.

In order to get the screen back to its previous condition we select **Zoom** again in the menu bar and the command line again displays

```
All/Centre/Extents/Previous/Windows/<Scale(X/XP)>:
```
Choose **Previous**, but you only need to type **P** at the command line and the screen will return to its previous form.

Note that if you use the **Zoom** feature more than once in succession and wish to obtain the original grid, you only need to select **Zoom** again at the command line, then type the letter **A** from the All option. You have complete control over the separate stages of magnification and this is important for accurate work.

Click on **Draw** in the menu bar, select **Line** and the command line displays `Command: _line From point:`. It is now a simple matter to click with the mouse anywhere on the screen for the start of a line and the instruction changes to `To point:`. Click again anywhere for the finish point and press <Enter>. The line is drawn and the next instruction can be given. Familiarise yourself with the changing instructions in the command area while you draw.

Note: when the **Ortho** button is highlighted in the toolbar, you will only be able to draw horizontal and vertical lines. Click the **Ortho** button off and lines at any angle can be drawn. The cursor will only move and snap into positions in increments of 2mm apart since this is the value you set in the Drawing Aids dialogue box. These adjustable values are set according to the precision you may require on the current drawing.

If you cancel the **Snap** feature by clicking the button, its highlight will go out and you can select any point on the drawing area. Toggling between **Snap** and **Ortho** will become part of the normal routine drawing process. You can clean the screen at any time by using the **Erase** command in the **Settings** menu. Two options are available to you here since you can click on a particular line or alternatively draw a square to completely enclose a

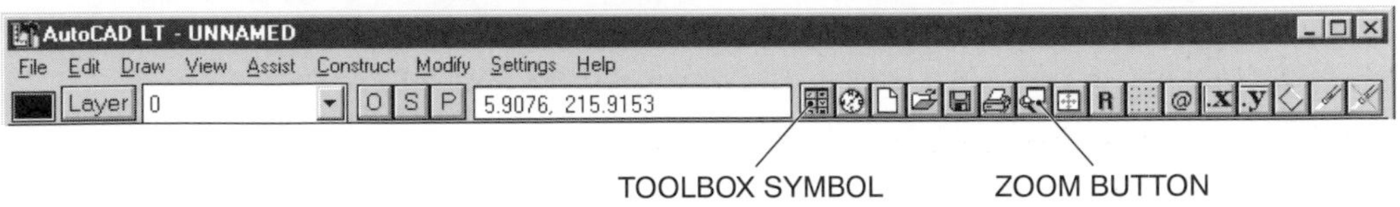

TOOLBOX SYMBOL   ZOOM BUTTON

Figure 1.6

number of items. Just click and then press <Enter>. This square could of course completely enclose the screen area. Repeated erasures may also leave the screen with many points left over from previous commands, and to clean these up you have the choice of selecting the **Redraw** option from the **View** pulldown menu or the easier method of typing the letter **R** at the command prompt, then press <Enter>.

All the operations so far have used the left-hand mouse button. When you were experimenting using the line command, you may have wanted to draw several lines one after the other in different positions. You will find it is not necessary to go back to the Toolbar each time. After drawing one line and clicking with the left mouse button to complete the command, the program will expect you to draw another line if you now click the right-hand mouse button. Any of the drawing commands can be repeated in this way. Just experiment to become familiar with the method.

The toolbox

When you turn the program on, the toolbar appears as shown in Fig. 1.6. The toolbox is a movable window containing symbols for commands used regularly. It can be positioned vertically at the left, vertically at the right, or horizontally by picking one of the bottom corners and moving to a more suitable position on the screen. It will be turned off in the fourth position. Toggle on the symbol in the toolbar to demonstrate the options. The contents of the toolbox and bar can easily be customised so that commands in regular use are rearranged to suit your particular draughting style.

Finished outlines

The outlines of all drawings consist of straight lines and curves of various lengths and radii. The essential details that the draughtsman needs to know are simply:

- the line lengths
- circle and arc radii
- the exact positions where the various combinations intersect.

It will also be necessary to decide where to start drawing on a blank sheet of paper. After commencing a drawing it is often the case that you change your mind. With manual draughting, a change of mind involves erasure. In CAD there is no problem since you can either change the paper size, change the drawing scale or move the drawing.

Exercise 1

Fig. 1.7 shows the layout of three pulley wheels with a continuous belt in a mechanism. The belt profile consists of three circular arcs and three straight lines. Lines AB, CD and EF are tangents to the pulley circles with tangency points A, B, C, D, E and F. Lines HA, HF, JB, JC, KD and KE are normals and the angle between each normal and tangent is 90°. These features are highlighted in the partly finished drawing in Fig. 1.8. Set up a new A4

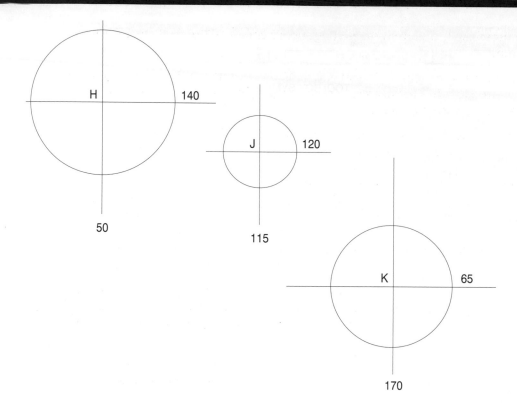

Figure 1.7

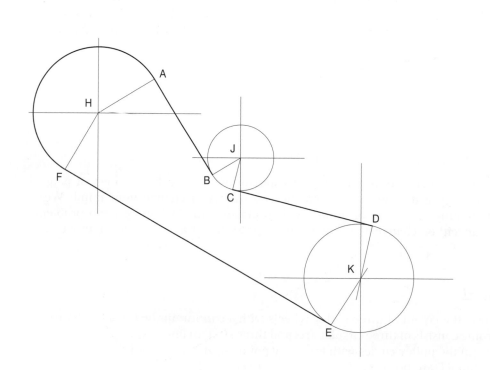

Figure 1.8

sheet in landscape style and from the Settings menu choose a 10 mm **Grid** and a **Snap** value of 1 mm.

There are many ways to input positional data in CAD and one of them is to use the co-ordinates feature, since it is the nearest thing to using a pencil and scale rule which most of us would use on a sheet of paper.

At the outset I would like to say that alternative methods each have their merits. However, from my experience of teaching the subject from the earliest days of CAD, I have found that it is far better to start to use the equipment with the same practical reasoning and questioning style associated with using pencil and paper. Having developed confidence and produced some sound results, it is then the time to apply sophisticated techniques, made possible by the computer and often applicable, where volume production and repetition is involved. It is also essential from the start to have complete command and control over the position of each line and feature. I hope you will follow the instructions closely and obtain an exact reproduction of each illustration in turn.

Complete Fig. 1.9 by drawing vertical lines where the X values are 50, 115 and 170. Add the horizontal lines with Y values of 140, 120 and 65. You will appreciate that the distances between these vertical lines correspond to the horizontal dimensions on the drawing of 65 and 55 mm.

Also, the horizontal lines are spaced to agree with the vertical distances of 55 and 20 mm. Providing they are long enough, the length of each line is unimportant, as finishing to a particular size can easily be undertaken later.

Continue by drawing the 60 diameter circle at centre H. Select **Draw** from the menu bar, then **Circle** and choose the first option of **Centre** and **Radius**. At the Command line the following options will appear: 3P/TTR/<Centre point>. Click over the intersection 140,50. The Command prompt will be Radius. Enter **30** and the circle is drawn.

If you now click the right-hand button on the mouse, the instructions will repeat themselves in the same order. Move the mouse and click over the next intersection at 120,115. The command line response will be: Radius<30>. This is not the value you

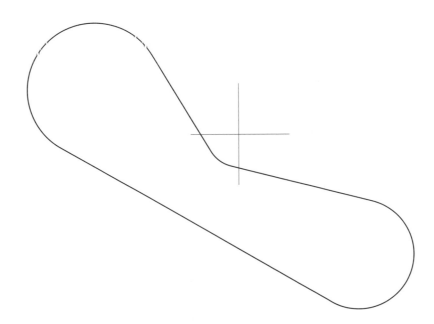

Figure 1.9

require so enter **15** and press <Enter>. Repeat the procedure at intersection 65,170 and enter the other value of 25. You will appreciate that for drawing the same size circles at different centre distances the left and right-hand mouse buttons can be used alternately.

The three tangent lines can be drawn automatically in turn. Disable the **Ortho** and **Snap** features. Select **Draw** from the menu bar. Choose **Line** followed by **Assist** in the menu bar. Select the **Object Snap** mode. The Running Object Snap dialogue box will now appear (Fig. 1.10).

The command line now reads `line From point: ddosnap`. Click on **Tangent** and press <Enter>. The command line changes to `line From point: ddosnap From point`. Click on one of the circle circumferences and then read `To point`. Click on the second circumference. The tangent line is drawn and having completed the operation the computer awaits the next instruction and shows simply `Command:`.

Notice the upper- and lowercase letters in the command space and the choice given to you. After making your selection, insert your option and press <Enter>.

To fix each tangency point turn off **Snap** and **Ortho**. Choose **Line** from the **Draw** menu, then **Object Snap** in the **Assist** menu. Click on the **Centre** option and press <Enter>. The command line now reads `line from point: ddosnap`. Click the left mouse button on the circle circumference. Notice that a line will now appear from the circle centre. Click again on the **Object Snap** button in the **Assist** menu, cancel previous **Centre** option and select **Perpendicular**. The command line now reads `To point: ddosnap`. Click the left mouse button on the tangent line followed by <Enter> and the normal will be drawn at right angles to the tangent. Repeat the process to establish the other five points of tangency.

We now have construction lines in place giving three tangents and the positions of six tangency points and it is necessary to finish our drawing to professional standards. BS 308 clearly states that drawings should be produced using two contrasting line thicknesses, thick and thin in the ratio 2:1. Professional manual draughtsmen generally use pens of thicknesses 0.35 and 0.7 mm. The Standard also lists the types of lines to be used for different applications and these are illustrated later.

Some experimentation may be required here as different plotters, printers, inks and papers give slightly different results. I normally use a line thickness of 0.4 mm for outlines

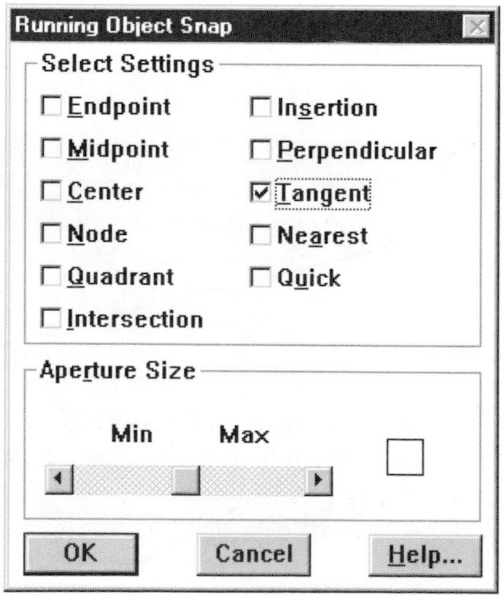

Figure 1.10

and leave the default thickness for centre and construction lines. My Canon BJ-230 ink-jet printer then gives acceptable linework.

Some parts of the circles are not required on the finished drawing but may easily be removed. Select the **Break** option in the **Modify** menu. At the command line the instruction reads Command: _break Select object.

We need to choose the circle to be broken, then select a small part of the circumference for the break. With a circle there are obviously two options. You can click on the circumference and then the position of the second click can be clockwise or anticlockwise in relation to the first point. Using one option close together and anticlockwise will remove just a small arc. The other option with the second point taken in the clockwise direction will remove the large arc. Just experiment here to prove the point.

To clean up the circle from centre H in Fig. 1.8 click on the circumference at about the 5 o'clock position and the circumference will appear in a dotted line. Click again anti-clockwise at about 4 o'clock and the small arc will disappear. We now have a broken circle. Using the BREAK command again click on the circumference at the 5 o'clock position and also on tangency point F, then repeat for the 4 o'clock position and tangency point A and the required arc will be complete.

Clean up the required arcs from the circles with centres J and K and the profile of the belt is defined, but not with the correct line thickness. The thickness of lines can be changed by clicking **Modify** in the menu bar and choosing **Edit Polyline**. The command line will now read Command: _pedit Select polyline:. We have three arcs and three straight lines to edit, so click on any one of them and the command line now reads:

```
Entity selected is not a polyline
Do you want to turn it into one?<Y>
```

The answer is 'yes' so press <Enter> or click the right-hand mouse button and the command line now offer many options as follows:

```
Close/Join/Width/Edit vertex/Fit/Spline/Decurve/Ltype gen/Undo/
eXit.
```

The first thing is to establish a new line thickness so type **W** on the keyboard and press <Enter>. Command line now reads Enter new width for all segments:. Type **0.4** and press <Enter>. The line thickness is increased and the list is repeated at the command line:

```
Close/Join/Width/Edit vertex/Fit/Spline/Decurve/Ltype gen/Undo/
eXit<X>
```

We have the result required, so just press <Enter> and the command line will clear and await the next request. Repeat the procedure for the other parts to complete the belt profile with a continous polyline 0.4 mm thick. Note again that by pressing the right-hand mouse button immediately after you pressed <Enter> that the Modify options were reinstated at the command line. Use of the right-hand button for repeat commands will improve your speed of working.

If you later wish to thicken up the profile of a complete thin circle and are using an early version of AutoCAD you may find when you select the Edit POLYLINE command that the software will not accept your entry. The secret here is to make a small break in the circle circumference and try again. You will be lucky this time, but the polyline is incomplete. After the thick line is drawn the command line repeats the following instructions:

```
Close/Join/Width/Edit/vertex/Fit/Spline/Decurve/Ltype gen/Undo/
eXit<X>
```

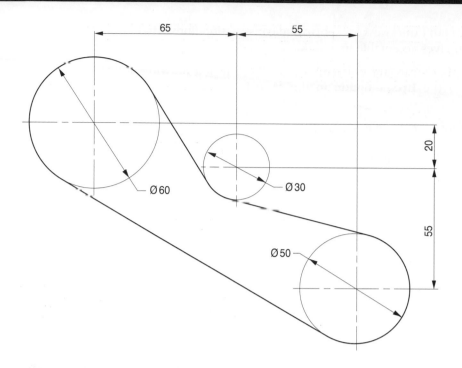

Figure 1.11

Type the letter **C** for close followed by pressing <Enter> and the circle will close with the required line thickness. Save your completed drawing as shown in Fig. 1.11.

You will find that some programs will not edit a complete circle, but if you require the thinner circle, as we do in this case to use as a dimension line, it can easily be redrawn afterwards. Edit the centre lines neatly as shown by using the BREAK command to remove small gaps as indicated. It is good practice to position the centre lines so that the centre itself is marked clearly by a cross. Try to make centre lines symmetrical. Use the BREAK command to clear unwanted lines. Save your solution fully dimensioned as shown in stage 3.

Before proceding, however, please refer to the general dimensioning notes on page 110.

Printing procedure

A print is now required of the completed pulley layout drawing and I assume a printer has already been installed and configured. Select the **Print/Plot** option from the **File** menu and the Plot Configuratión dialogue box will appear on the screen (Fig. 1.12). Click at the top left corner on the **Print/Plot Setup Default Selection** option and the Print Setup dialogue box appears and confirms the printer specification (Fig. 1.13).

My printer is a Canon BJ-230 using standard A4 size paper. Check that your printer is listed and the other details are as above. Click **OK**. The Print/Plot Setup box returns. We need a copy of the screen image so select **Display**. Check the paper size box. Click on the **Rotation** and **Origin** button and the Plot Rotation and Origin dialogue box shown in Fig. 1.14 appears.

The drawing has been produced with the paper turned through 90°, so click on the plot rotation setting of **90** and **OK**. The first Plot Configuration box returns. To ensure that our print is the maximum size for the paper area available, click on the **Scaled to Fit** box in the bottom right-hand corner. Click on the **Preview** and **Full** buttons and the print details will

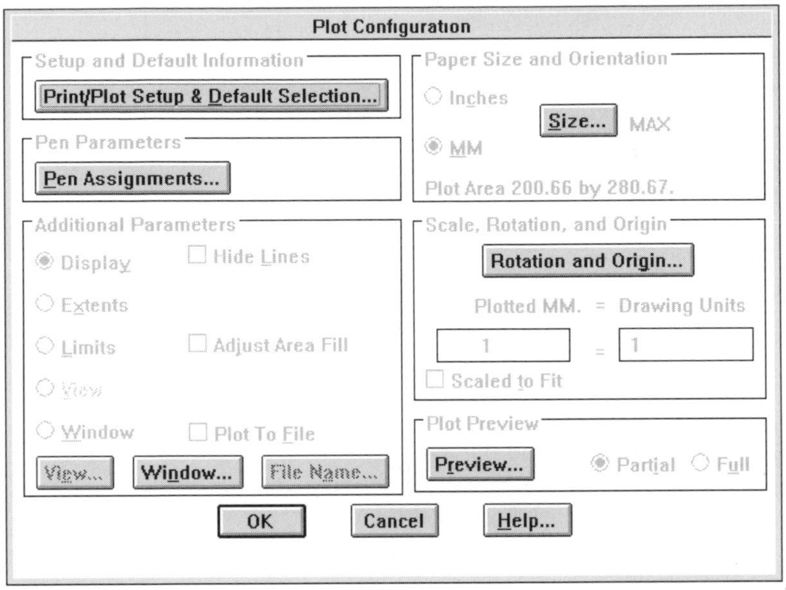

Plot Configuration

Setup and Default Information
Print/Plot Setup & Default Selection...

Pen Parameters
Pen Assignments...

Additional Parameters
⦿ Display   ☐ Hide Lines
○ Extents
○ Limits   ☐ Adjust Area Fill
○ View
○ Window   ☐ Plot To File
View...   Window...   File Name...

Paper Size and Orientation
○ Inches   Size...   MAX
⦿ MM
Plot Area 200.66 by 280.67.

Scale, Rotation, and Origin
Rotation and Origin...

Plotted MM. = Drawing Units
        1   =   1
☐ Scaled to Fit

Plot Preview
Preview...   ⦿ Partial ○ Full

OK   Cancel   Help...

Figure 1.12

appear. Now click on the **OK** button at the bottom of the box if all is satisfactory. A note will appear in the command box if, for any reason, the printer is not ready.

ARRAY command

An interesting and often entertaining feature is the ARRAY command in the Construct menu. This can be used, for example, to arrange parts symmetrically about a given centre. As an introduction to this option save a copy of your drawing of Fig. 1.9 under a separate drawing number.

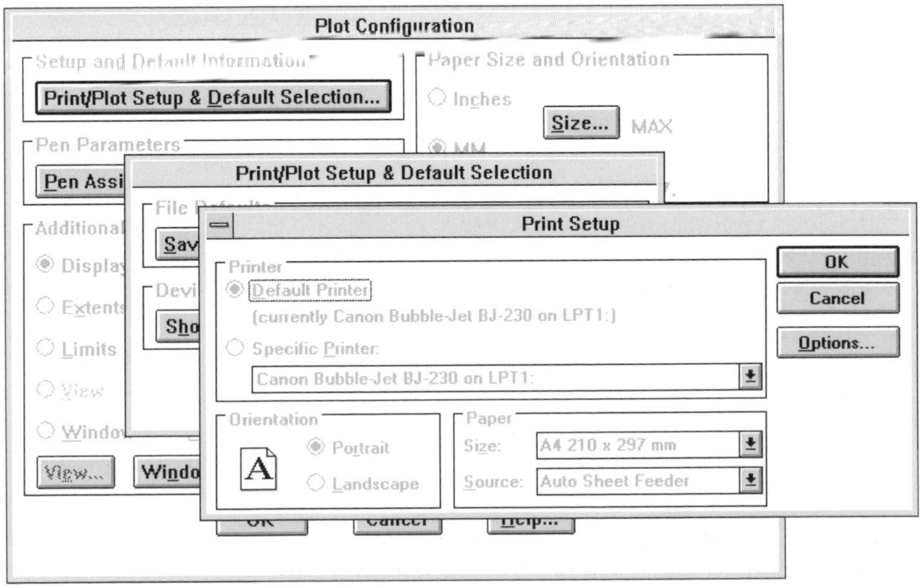

Plot Configuration

Setup and Default Information
Print/Plot Setup & Default Selection...

Pen Parameters
Pen Assi

Additional
⦿ Display
○ Extents
○ Limits
○ View
○ Window
View...   Windo

Paper Size and Orientation
○ Inches   Size...   MAX
⦿ MM

Print/Plot Setup & Default Selection

File Defaults
Sav

Devi
Sho

Print Setup

Printer
⦿ Default Printer
   (currently Canon Bubble-Jet BJ-230 on LPT1:)
○ Specific Printer:
   Canon Bubble-Jet BJ-230 on LPT1:

Orientation
A   ⦿ Portrait
     ○ Landscape

Paper
Size: A4 210 x 297 mm
Source: Auto Sheet Feeder

OK
Cancel
Options...

Figure 1.13

Plot Configuration

Setup and Default Information

Print/Plot Setup & Default Selection...

Paper Size and Orientation

○ Inches

○ MM

Size... MAX

Pen Parameters

Pen Assig...

Plot Rotation and Origin

Plot Rotation

○ 0   ◉ 90   ○ 180   ○ 270

Plot Origin

X Origin: 0.00   Y Origin: 0.00

OK   Cancel

Additional

◉ Display

○ Extents

○ Limits

○ View

○ Window

Drawing Units

View...   Window...   File Name...   Preview...   ◉ Partial ○ Full

OK   Cancel   Help...

Figure 1.14

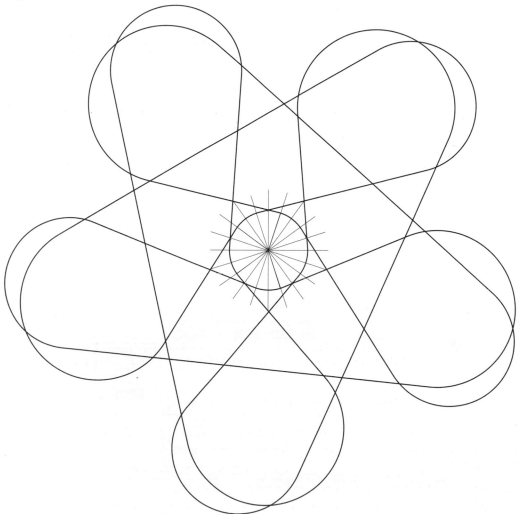

Figure 1.15

Select **Array** in the **Construct** menu and the command line reads

```
Command: _array
Select objects:
```

Click on the drawing at the bottom left-hand corner and then again at the top right-hand corner so that the drawing is enclosed by a box. The command line reads `Select objects: Other corner: 16 found`. The drawing is highlighted so click and press <Enter>. The command line now reads `Rectangular or Polar array (R/P) <R>:`. Type **P** and press <Enter>. The command line answers `Centre point of array:`. Click on the intersection of the centre lines that positioned the axis through wheel J. Press <Enter> and the command line responds `Number of items:`. Let us assume that we would like five parts to our illustration, so type **5** then press <Enter> and the command line replies `Angle to fill (+=ccw, -=cw) <360>:`. We require our five parts equally spaced around 360° so simply press <Enter> again and the drawing shown in Figure 1.15 will appear.

Fillet applications

The FILLET feature is useful for joining lines and arcs automatically with an arc of a specified radius. Fig. 1.16 shows three pairs of lines which I have drawn at random angles and with random radii.

I want to join the lines with a fillet radii of, for example, 14 mm and have spaced the pairs of lines so that such a size of radii will fit. If it does not do so because the lines are not far enough apart to accommodate it, the computer will reject the fillet request and this message will appear:

```
Radius is too large
'Invalid'
Command:
```

To set the fillet radii choose **Fillet** from the **Construct** menu and the command line will read:

```
Command: _fillet Polyline/Radius/<Select first object>:
```
Type **R** and the line will change to `Enter fillet radius <2.0000>:` 14. Press <Enter> and the line returns to `Command:`

Note that the value of 2.0000 above was the previous value used. The value of 14 mm is the current set value for the fillet radii and will remain until we modify it. Choose **Fillet** again from the **Construct** menu and the line returns to

```
Command: _fillet Polyline/Radius/<Select first object>:
```

Click on one line and the command line changes to `Select second object:`. Click on the second line of the pair and the fillet is drawn. Repeat the procedure for the other pairs of lines.

The situation now is that we have a continuous thin line but would like to define the tangency points and the fillet arc centre. Finally we need to thicken up the profile for a finished outline. From the **Modify** menu choose **Edit Polyline** as previously described and adjust the line width to 0.4 mm. Repeat this again for the other pairs.

Turn off the **Ortho** and **Snap** features since the line we are about to draw will come from an exact point which will not coincide with points on our grid. In order to find the centre points for the fillet radii we can again use the **Draw** menu, select **Line**, then click on **Assist** and in the Running Object Snap box make sure there are no other entries and select **Centre**. The command line reads

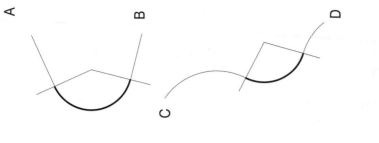
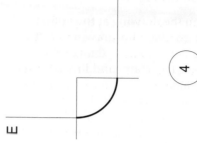
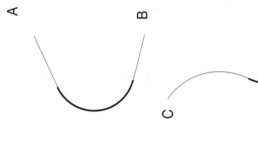
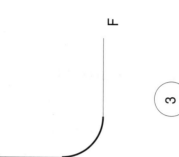

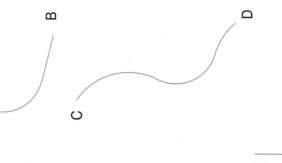
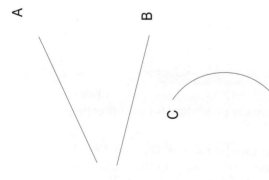
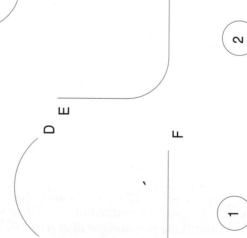

Figure 1.16

```
Command: _line From point: '_ddosnap From point:
```

Click on the fillet curve, the centre point is indicated and a line will appear. The command line will read To point.

The tangency point can clearly be seen at the end of the thickened line, so click the left button again just beyond it, having ensured that your line passes through the end of the fillet radius. Click again on the right-hand button and you can repeat the procedure for the line through the other tangency point. Repeat for the other examples as shown.

You may well find that before proceeding with this part that it will be advantageous to use the ZOOM feature for each construction to increase the size of the drawing and improved accuracy especially as the SNAP feature has been disabled.

I have found that in especially geometrical work it is a good habit to check accuracy with ZOOM control. Where work progresses in steps, each dependent on the previous move, it is not good practice to carry forward small errors of inaccuracy in input data which could have been avoided. So if you do make errors, go back and correct them as you will certainly have regrets later. Never forget that if you are working professionally that your work can be easily checked by your superiors. It is also a fact that the computer can record the day and time details when entries are made. Don't forget either to return to using the SNAP feature when possible.

Part of a cover for an air vent is illustrated in Fig. 1.17. The example is given to introduce the COPY and MIRROR features. Ensure that ORTHO and SNAP are activated.

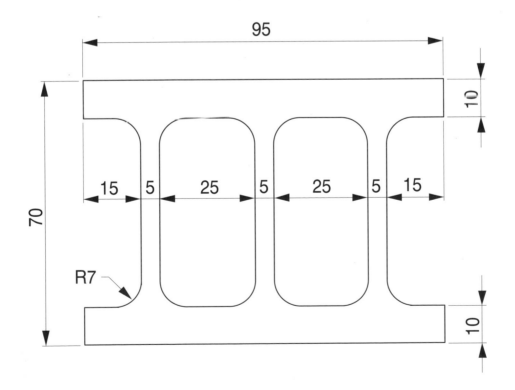

Figure 1.17

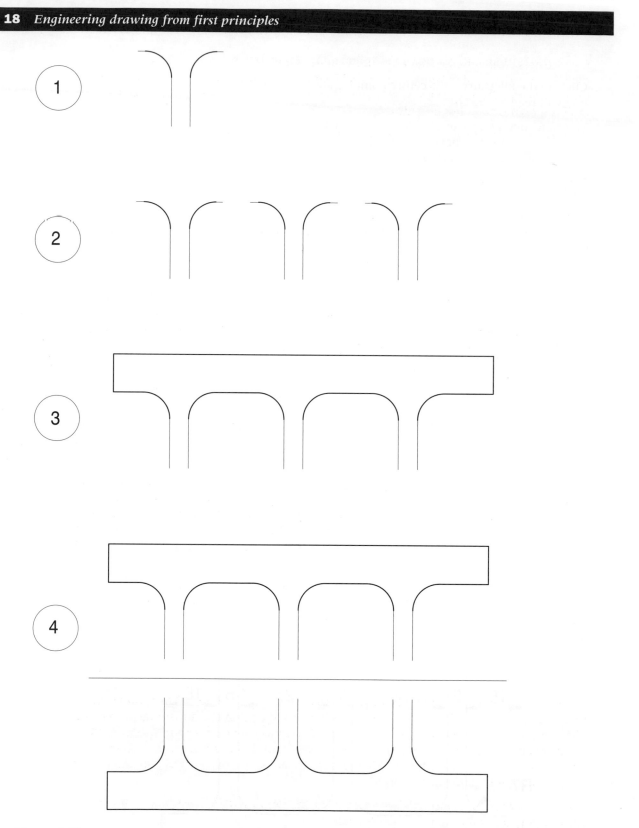

Figure 1.18

You will see that the cover mainly consists of six pairs of fillet radii and it is not necessary to draw all of these as separate parts. Set out stage 1 (Fig. 1.18) with two vertical lines 5 mm apart and a short horizontal line. Use the FILLET command reset to 7 mm and insert the two arcs. You will find that after the first arc is inserted that you will have to redraw the short horizontal line because the FILLET command trims the construction lines back to the tangency points. In many drawings this is a very useful feature. Change the two radii to polylines 0.4 mm in width.

For stage 2 choose the **Copy** command from the **Construct** menu and the command line reads

```
Command: _copy
Select objects:
```

You are now required to draw a small box to enclose the drawing in stage 1. Click the mouse over the bottom left-hand corner and again over the top right-hand corner and the command line now reads `Select objects: Other corner: 6 found`. Note that the contents of the box have been highlighted.
Press <Enter> and you now need a datum line to indicate where the COPY starts from and also where it is required to be redrawn. The instruction at the command line changes to `<Base point or displacement>/Multiple:`. Click on the vertical line at the side of the left-hand fillet radii as our datum, then note from the dimensions of the required drawing that we need to move the contents of the box 30 mm to the right. The command line now reads

```
Select objects: <Base point or displacement>/Multiple: Second
point of displacement
```

Having clicked on our datum line, just move it the 30 mm to the right. Click again to enter the movement with the left-hand mouse button. We need two copies, so click again with the right-hand button and repeat the operation.

You will see that the co-ordinate display indicates your 30 mm moves and also the angle with complete accuracy, and this is how we can control positioning to perfection when important datum parts of drawings are located on grid lines. Furthermore, I always try to start drawings with centre lines or datum edges on 10 mm grid lines just for convenience and ease of marking out. The arithmetic is then easy to observe and check.

For stage 3 add the polylines as shown in Fig 1.18. In stage 4 we need to repeat the top profile upside down so that both halves are equally spaced 35 mm above and below the mirror line. Select **Construct** from the menu bar, then **Mirror** and the command box display reads

```
Command: _mirror
Select objects:
```

Draw a rectangular box around the illustration for stage 3 and <Enter>. The command line reads `First point of mirror line:`. Click on one point on the horizontal mirror line and the command line reads `First point of mirror line: Second point:`. Click again on another point on the mirror line and the command line answers `Delete old objects?<N>`. We haven't any old objects to delete so press <Enter> and the mirror image is drawn to perfection.

Nothing now remains but to delete the mirror line with the **Erase** command, insert the vertical polylines and the drawing is complete.

General drawing recommendations

The exercises so far have introduced many drawing commands to get a feeling for the work and all examples have been drawn on a single drawing layer. Various branches of industry and commerce have developed established practices and standards and your local Reference Library is an ideal source of information. The following notes are included here for background reading and information.

BS and ISO Drawing Standards cater for all drawings no matter how they are prepared. The Standards do not differentiate between drawings produced by electronic means, ink tracings or pencil drawings.

CAD linework is black and dense and what you see on the screen is exactly what the printer or plotter will reproduce. There are no problems with line density and variable

Table 1.1

Line	Description	Application
A	Continuous thick	A1 Visible outlines A2 Visible edges
B	Continuous thin	B1 Imaginary lines of intersection B2 Dimension lines B3 Projection lines B4 Leader lines B5 Hatching B6 Outlines of revolved sections B7 Short centre lines
C D	Continuous thin irregular Continuous thin straight with zigzags	[1]C1 Limits of partial or interrupted views and sections, if the limit is not an axis [2]D1 Limits of partial or interrupted views and sections, if the limit is not an axis
E F	Dashed thick Dashed thin[3]	E1 Hidden outlines E2 Hidden edges F1 Hidden outlines F2 Hidden edges
G	Chain thin	G1 Centre lines G2 Lines of symmetry G3 Trajectories and loci G4 Pitch lines and pitch circles
H	Chain thin, thick at ends and changes of direction	H1 Cutting planes
J	Chain thick	J1 Indication of lines or surfaces to which a special requirement applies (drawn adjacent to surface)
K	Chain thin double dashed	K1 Outlines and edges of adjacent parts K2 Outlines and edges of alternative and extreme positions of movable parts K3 Centroidal lines K4 Initial outlines prior to forming [4]K5 Parts situated in front of a cutting plane K6 Bend lines on developed blanks or patterns

Notes

1. The lengths of the long dashes shown for lines G, H, J and K are not necessarily typical due to the confines of the space available.
2. This type of line is suited for production of drawings by machines.
3. The thin F type line is more common in the UK, but on any one drawing or set of drawings only one type of dashed line should be used.
4. Included in ISO 128–1982 and used mainly in the building industry.

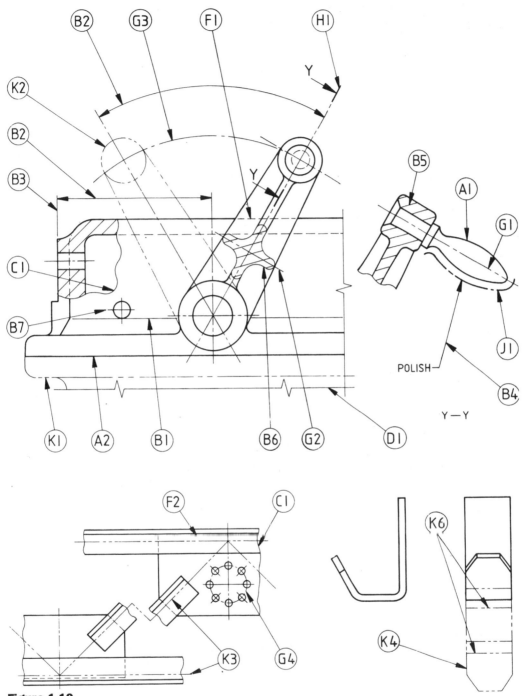

Figure 1.19

width from pencil wear which the manual draughtsman needed to overcome to produce quality work. Professional ink tracings gave a much-improved finish but modifications to ink and pencil drawings often presented problems on the drawing surface. Only two line thicknesses are required by current British Standards: thick and thin in the ratio of 2:1. The original illustrations for this book have been drawn with a polyline width of 0.4 mm which is adequate to emphasise the thick outlines.

I use ordinary 80 gsm budget A4 copier paper on a Canon BJ-230 inkjet printer.

Inks evaporate and papers absorb at different rates, so some experimentation may be necessary with different combinations of papers and printers to obtain contrasting line thicknesses. The drawings in this book are designed to fit A4 size sheets and I hope you will copy them to improve your draughting knowledge and efficiency. Perhaps the evidence of your ability will come in handy when seeking employment.

The British Standard linework recommendations are shown in Table 1.1 by kind permission of The British Standards Institution. Applications of the various linetypes are shown in Fig. 1.19.

Coinciding lines

It is often the case that two or more lines will coincide, and the following order of priority is recommended by BS 308 (Table 1.1).

1. visible outlines and edges (types A);
2. hidden outlines and edges (types E or F):
3. cutting planes (type H1);
4. centre lines, etc. (types G and B7);
5. outlines and edges of adjacent parts, etc. (types K);
6. projection lines (type B3).

Adjacent outlines of assembled parts should coincide. Examples of the priorities of coinciding lines are shown in Fig. 1.20, and all of the illustrations which follow in the book are applications of these principles.

The quality of finished linework can be enhanced by attention to the following small details. Generally for most applications, dashed lines should consist of a line about 3 mm long with a 2 mm space and CAD can give lines of these proportions.

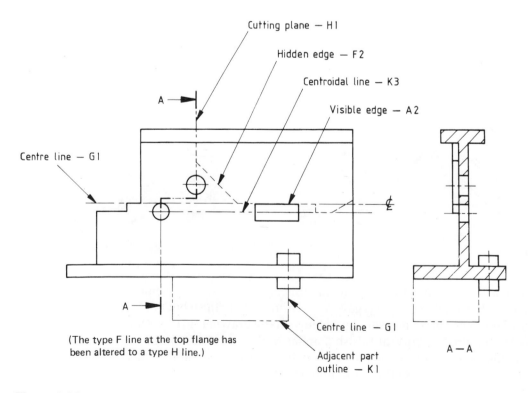

Cutting plane — H1
Hidden edge — F2
Centroidal line — K3
Visible edge — A2
Centre line — G1
A
A
Centre line — G1
(The type F line at the top flange has been altered to a type H line.)
Adjacent part outline — K1
A — A

Figure 1.20

Table 1.2

Applications	Drawing sheet size	Minimum character height (mm)
Drawing numbers etc.	A0, A1, A2, A3	7
Dimensions and notes	A4	5
	A0	3.5
	A1, A2, A3, A4	2.5

Dashed lines start and finish with a dash in contact with connecting outlines or other hidden lines. Dashes meet at corners and junctions where changes in direction occur. A dashed line joining a curve should start with a dash at the tangency point.

Chain lines are drawn with long and short alternate dashes and start and finish with a long dash. Generally the proportions of chain and centre lines should be adjusted so that they appear to be balanced and symmetrical if possible. Centre lines extend a short distance beyond an outline. If centre lines cross they should be arranged to intersect on a line rather than in the space.

Where centre lines define a circle centre, the centre point should be defined by crossing lines. Where a succession of holes are drawn, the centre lines appearance can often be improved by drawing one hole with its centre lines and using the COPY or ARRAY feature to repeat as required.

Minimum character height proportions are given in Table 1.2. All dimensions and notes have been drawn with a minimum character height of 3.5 mm and this is meant to take into account the fact that some drawings need to be reduced in size to fit a book page. Roman Duplex is the text style selected with a width factor of 0.9. This width factor reduces the width of notes by about 10 per cent when compared with a width factor of 1. Please also refer to Chapter 5 regarding text applications.

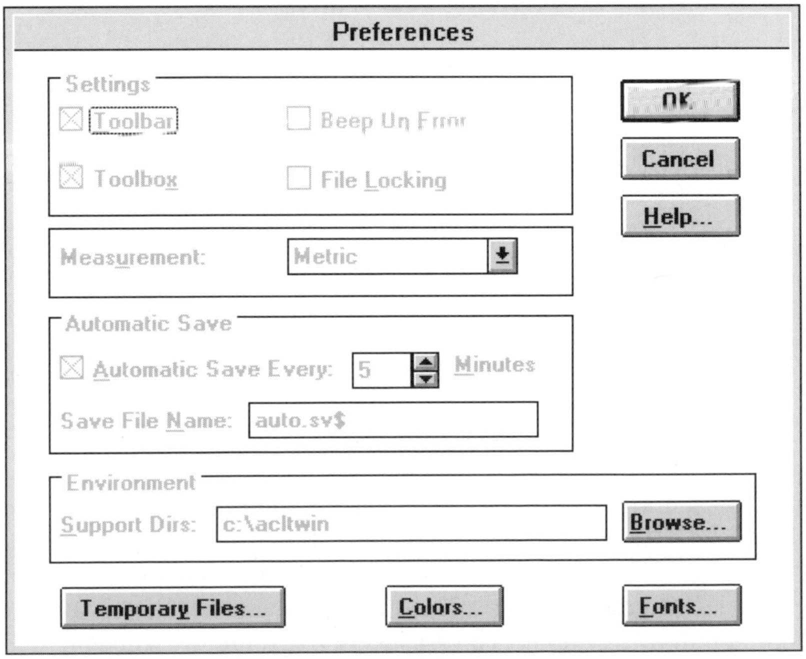

Figure 1.21

Table 1.3 Recommended abbreviations for drawing notes

Term	Abbreviation or symbol
Across flats	A/F
Assembly	ASSY
Centres	CRS
Centre line	₡ or CL
Chamfered	CHAM
Cheese head	CH HD
Countersunk	CSK
Countersunk head	CSK HD
Counterbore	C'BORE
Cylinder or cylindrical	CYL
Diameter (in a note)	DIA
Diameter (preceding a dimension)	Ø
Drawing	DRG
External	EXT
Figure	FIG.
Hexagon	HEX
Hexagon head	HEX HD
Insulated or insulation	INSUL
Internal	INT
Left hand	LH
Long	LG
Material	MATL
Maximum	MAX
Minimum	MIN
Number	NO.
Pattern number	PATT NO,
Pitch circle diameter	PCD
Radius (preceding a dimension, capital letter only)	R
Required	REQD
Right hand	RH
Round head	RD HD
Screwed	SCR
Sheet	SH
Sketch	SK
Specification	SPEC
Spherical diameter (preceding a dimension)	SØ
Spherical radius (preceding a dimension (ϕ)	SR
Spotface	S'FACE
Square (in a note)	SQ
Square (preceding a dimension)	☐ or ☒
Standard	STD
Undercut	U'CUT
Volume	VOL
Weight	WT

For convenience on A4 sheet drawings I use a standard arrow 4 mm long by 2 mm wide for dimensions and leader lines. If larger arrows are required I use the SCALE feature on the Modify menu and choose a value between 1 and 2. This not only increases the size of the arrow but keeps the length and width to the same proportions. Table 1.2 gives a minimum value of 5 mm for drawing numbers on A4 sheets. It is therefore convenient to keep the same text style, and whenever larger numbers and letters are required simply use the scale value greater than 1. It is time consuming to keep on changing text styles.

Capital letters are preferred on technical drawings. Lowercase letters are only used where they form parts of symbols or abbreviations. Table 1.3 shows the recommended list of abbreviations used for drawing notes.

Preferences

The File menu contains an option which allows you to make certain choices with regard to your style of working. Select **Preferences** in the **File** menu at the top of the screen and the dialogue box shown in Fig. 1.21 will appear.

Settings

Each of the drop-down menus is actuated by clicking on the Toolbar. The Toolbox is a handy feature since it can be turned on or off at any time or repositioned in different parts of the screen while you work. It can also be personalised with a choice of symbols. I suggest that you leave both on. The use of Beep On Error is optional.

File locking is not required unless you are part of a network system. This feature locks files to prevent other operators trying to use the same file as yourself. Computers can be linked together where several draughtsmen are working on the same contract.

Select **Metric** as the measurements unit.

The AUTOMATIC SAVE feature will save your work at an agreed time interval and protect it from loss in the event of power failure or other accidental calamities; it is very useful.

The COLORS and FONTS buttons allow you to change the screen background colours and textual details. A SYSTEM COLORS button allows a return to default settings after experimentation.

Backup files

I strongly recommend the good practice of regularly making backup files as a safeguard against loss if your hard disk fails. The time interval between backups will depend on your frequency of working. Reference to the section on Backing Up Your Files in the *Microsoft Windows and MS-DOS Users Guide* is essential reading.

Linetypes and layers

You will note from Table 1.1 that two thickness of lines are used in engineering drawing and that several different linetypes are required. CAD provides layers which are transparent overlays and the draughtsman chooses separate layers for each of the different linetypes needed.

The first job is to set up the layers. Click the **Layer** button on the left side of the menu bar and the Layer Control dialogue box appears (Fig. 1.22).

To draw most of the exercises in this book requires the use of four layers, since the drawings involve the following activities.

1. setting out centre lines to position component details;
2. setting out construction lines for geometric work;
3. inserting hidden details;
4. lining in finished outlines.

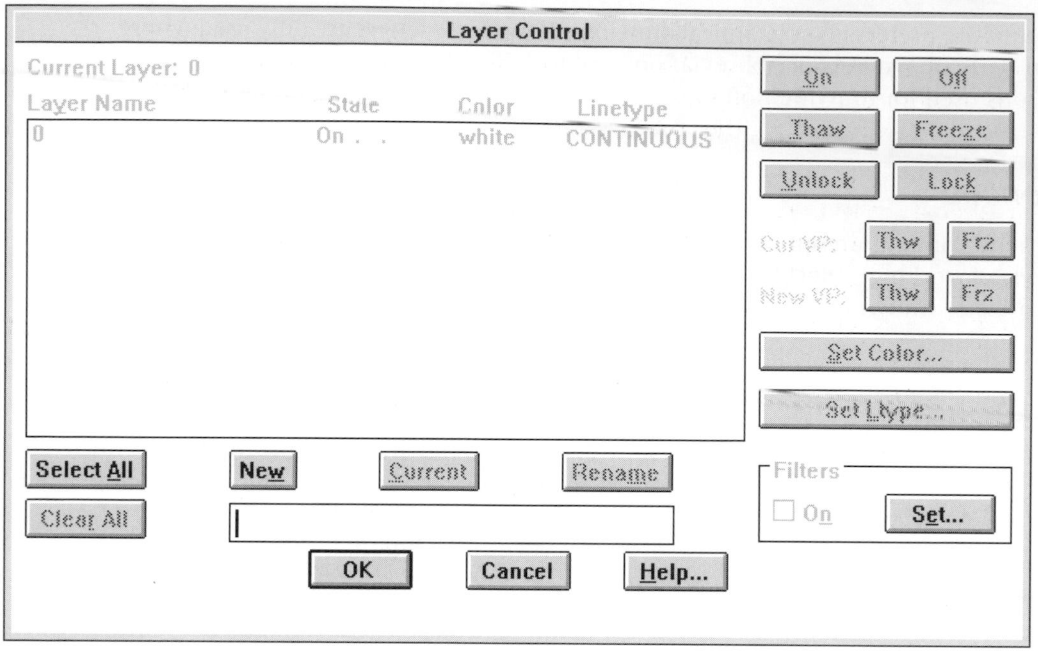

Figure 1.22

You can of course easily add others if necessary.

When **Line** is selected from the **Draw** menu on the first layer a thin chain line can be drawn. Constructional details with thin continuous lines are drawn on the second layer. Hidden detail is shown by thin dashed lines. The fourth layer is used for thick finished outlines.

Colours can be selected for each different line style. As well as making it easier to interpret the linework, it is also a pleasure to construct it in interesting colours.

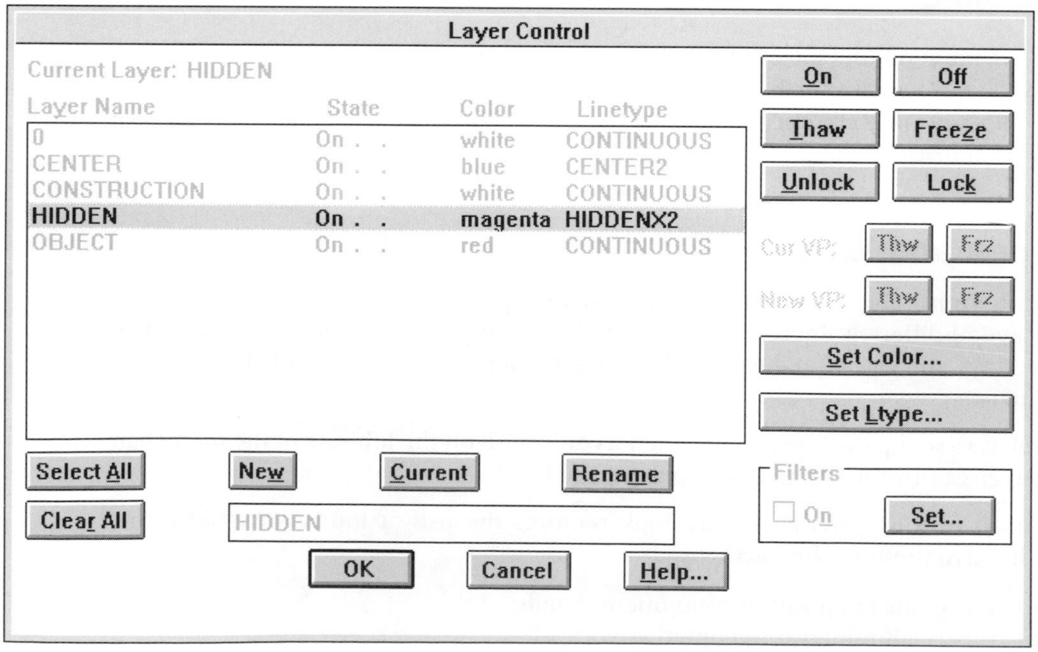

Figure 1.23

We can always add to the list but basically four linetypes will suffice at this stage, namely 'center', 'construction', 'hidden' and 'object' and each will have a distinctive colour. We do not use the default layer zero, so in the blank name input box at the bottom type the words **center**, **construction**, **hidden**, **object** and click on **New**.

The layer names will appear in the sequence above but all colours will be quoted as 'white' and all lines 'continuous'. These need to be edited. Click on the layer names in turn, then **Set Color** and **Set Ltype** and edit the details as indicated on Fig. 1.23.

A variety of broken lines are provided by the software and also a BREAK command which permits you to insert your own breaks conveniently in ordinary continuous lines if the proportions of the given lines do not suit your style. The option is also available to set up your own selection. Similarly, chain lines are required for hidden details and a standard selection is supplied. I choose my lines to give a 3 mm line followed by a 2 mm space. Again, you can design your own if you wish.

Chapter 2

Geometrical applications

This chapter introduces you to:

- Dimensioned tangency examples.
- Tangency points and intersecting arcs.
- Polar and Rectangular arrays.
- Ellipse construction and properties.
- Scale factors.
- Wire parts.
- Symmetrical layouts.
- Double lines and arcs.
- Wire springs.
- Setting out exact rotation angles using the Gear Wheel example.
- Gear Wheel terminology.
- UNDO, REDO, BREAK, EXTEND and TRIM features.
- Loci.
- Archimedian Spiral.
- Cylindrical and conical helix.

Tangency

In order to gain experience and proficiency, it would be desirable after completing each of the following drawing exercises to add the dimensions which are provided. Please accept this as a desirable form of 'on the job' training. Continuous practice is necessary to improve speed, familiarity and overall efficiency.

The dimensions follow the procedures and practices recommended in British Standards. It is essential to show accurate and unambiguous dimensional details.

Dimensioning is a very important part of engineering drawing and Chapter 5 details other aspects not illustrated in these examples. Please therefore refer to this information while you make progress in copying the applications throughout the book. The following examples emphasise where lines and curves join, so that the exact length and position of each small part can be understood.

Light bulb

Tangency points for the position of joins between intersecting arcs may be found by constructing lines between the arc centres (Fig. 2.1). Set out the given basic features and determine the positions of tangency points X and Y using the construction lines indicated from centres P, Q and R. Line in the profile with a polyline of 0.4 mm width and add the dimensions to complete the detail.

The lamp profile can also be drawn by positioning part of the R50 and R80 radii as shown

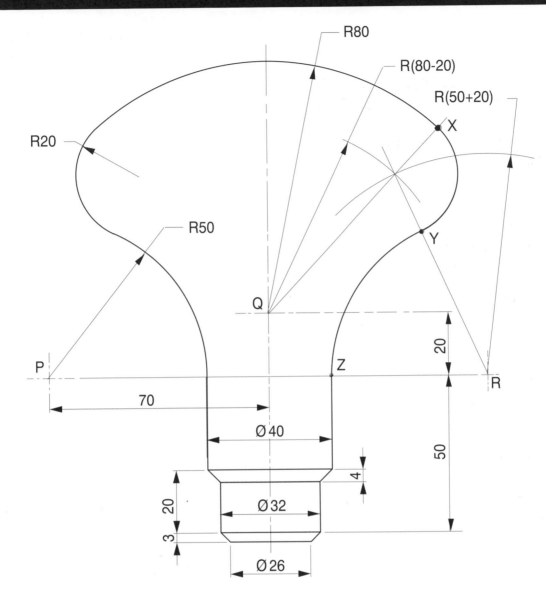

Figure 2.1

in Fig. 2.2 and then using the FILLET feature previously described to insert the R20 radius. If the arc resulting from the use of the FILLET feature is the first to be modified into a polyline, it will then be possible to define the tangency points at each end. Clean up surplus construction lines from the other two arcs and add the lower details shown. Use the MIRROR feature described in Chapter 1 by enclosing your profile in a box from A to B and clicking twice on the mirror line.

Film reel

The dimensions of one-quarter of a symmetrical film reel are shown in Fig. 2.3. Set out the profile shown and insert the four corner radii using the fillet feature. Use the MIRROR command about the OY axis to obtain half of the reel and again about the horizontal axis to complete the drawing (Fig 2.4).

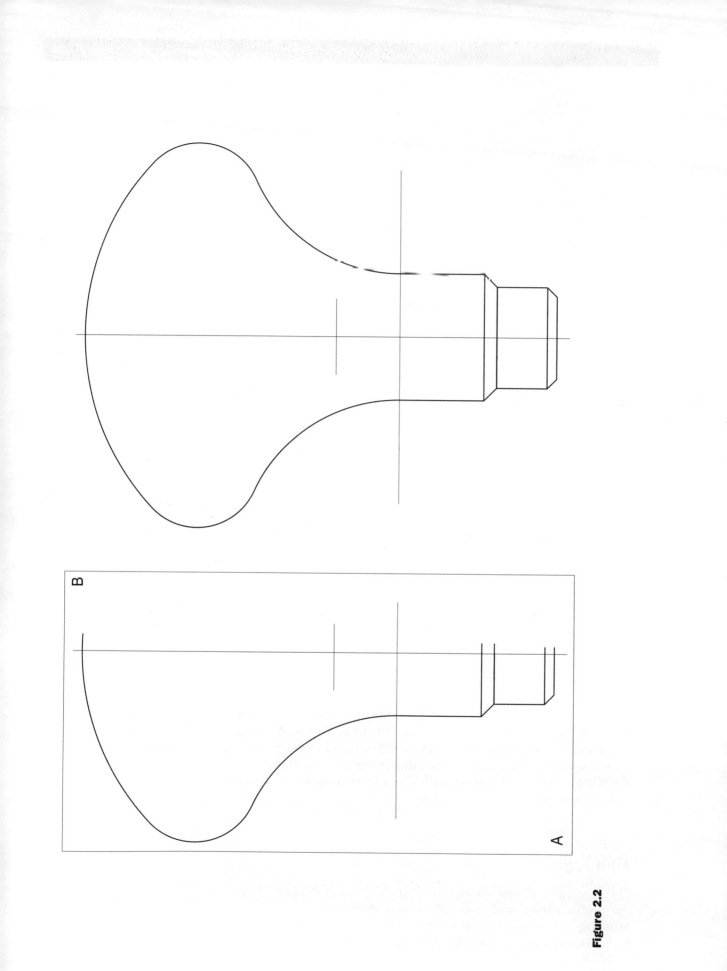

Figure 2.2

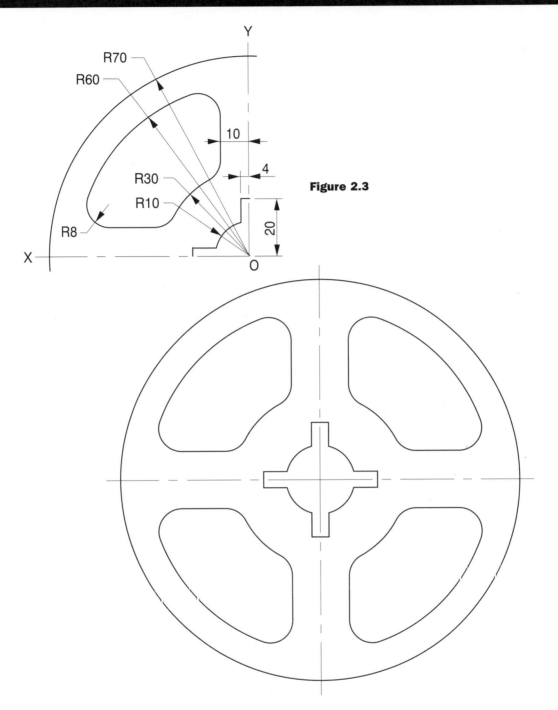

Figure 2.3

Figure 2.4

Bottle opener

Turn the drawing sheet into landscape style and copy the bottle opener (Fig. 2.5). Note that the R90 radii can be inserted using the FILLET feature. Add the dimensions to your solution.

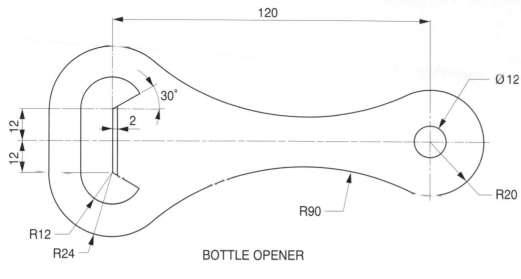

Figure 2.5

Restaurant sign

The example requires you to change the paper settings to a landscape-style drawing (Fig. 2.6). It is also necessary to establish the position of the centres and tangency points in connection with the R15 arcs. Change the SNAP settings temporarily to fix the centres for the eye.

Metal worker's dolly

Draw the profile of a metal worker's dolly shown in Fig. 2.7. The curve XYZ is a semi-ellipse. The dolly is supported by the taper. Thin sheet metal may be hammered into shape on the top surface.

Telephone receiver

Commence drawing the outline of the telephone receiver in Fig. 2.8 by setting out the major axes and positioning the two centres marked A and B. You may find it beneficial to draw the profile in simple construction lines and then modify the outlines into polylines. Use the FILLET command when appropriate and where the program automatically trims some construction lines, it is a simple matter to reinsert them again, if essential, before proceeding.

The example requires you to establish the centre of the R300 arc from the centres A and B, and it depends where you start the drawing on the sheet, but it may be necessary to reposition the illustration before printing. Remember to use the SNAP feature whenever possible. Check the position of all tangency points so that you obtain smooth accurate curves.

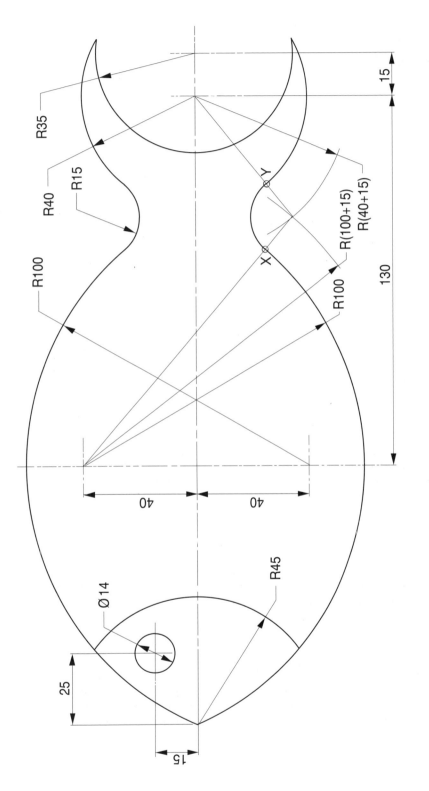

RESTAURANT SIGN

Figure 2.6

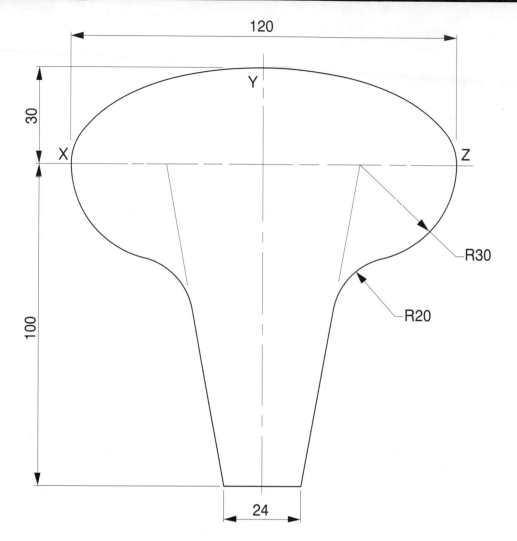

Figure 2.7

Plane handle

The plane handle in Fig. 2.9 is another problem involving establishing the position of arc centres and tangency points. Start by inserting construction lines to locate the centres of the two R18 arcs. Clean up unwanted lines after positioning tangency points and modify the outline into polylines.

Earthenware vase

The dimensions of a vase are given in Fig. 2.10. Copy the profile and add the dimensions. Start by drawing the R50 radius and construct the intersecting arcs to locate the centres for the other two radii.

Dipstick

Figure 2.11 shows a dipstick which is used to indicate the depth of oil in an engine. The stick is calibrated with five equal divisions between points A and B and you will note from

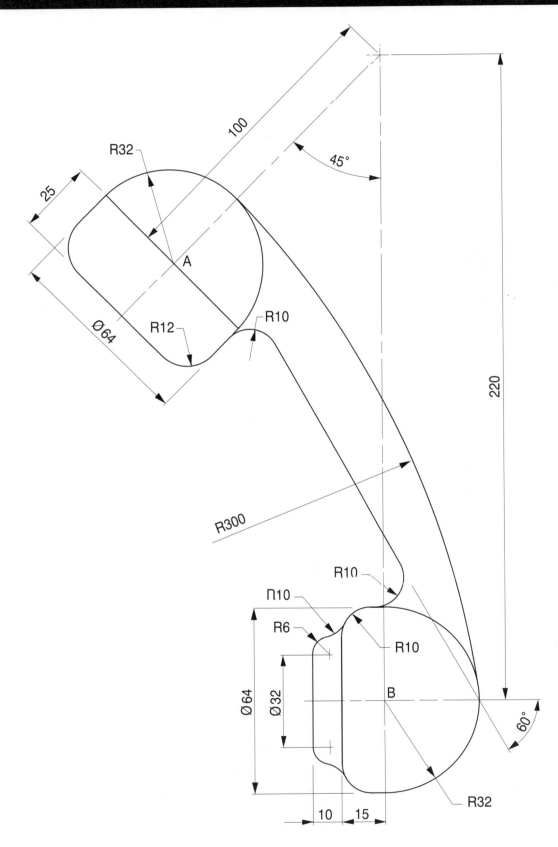

Figure 2.8

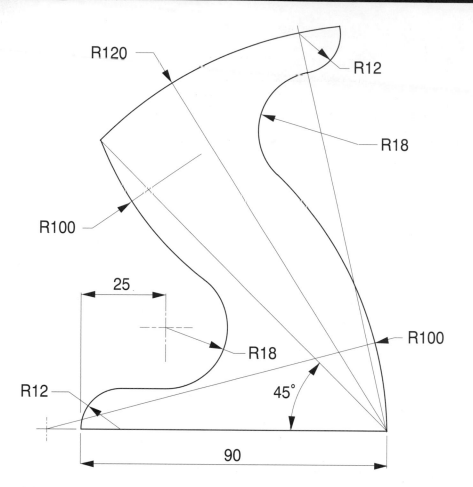

Figure 2.9

the dimensions given that the length of 107 is not readily divisible by 5. The construction here is one which uses the principle of similar triangles.

Draw vertical lines from points A and B and from a point O draw an arc with a radius of 110 mm which is divisible by 5. The line OY is now marked off using arcs of radii R22, R44, R66 and R88. The vertical lines through the intersections give the required calibration points.

Line in the outline with polylines 0.4 mm wide. The dipstick can be drawn using the FILLET option for the 80 mm radius and the rectangular array for the graduations. The geometrical method is given here for revision purposes.

Screwdriver

Determine the positions of all the arc centres and draw the profile of the screwdriver shown in Fig. 2.12.

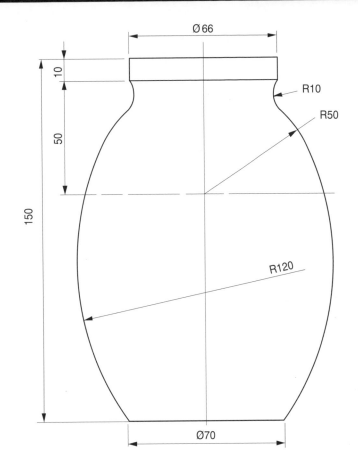

Figure 2.10

Propeller

Figure 2.13 shows one blade from a propeller. Draw the dimensioned profile. Set out the centres of the R30 and R50 arcs. From the centre of the R30 arc draw a line 15° from the horizontal to establish the tangency point and from it the line at 15° to the vertical. The fillet feature will complete the construction by inserting the R8 and R20 arcs. Clean up the construction lines and use EDIT POLYLINE to give the required outline. Save the drawing on a separate file. The intention is to reduce the size of the profile by using the **Scale** feature in the **Modify** menu. The command line will read

```
Command: _scale
Select objects:
```

Draw a box to surround the component and press <Enter>. The command line changes to

```
Select objects: Other corner: 35 found
Base point
```

Choose the centre of the circular boss and the command line reads <Scale factor>/ Reference:. We want to reduce the drawing to half size, so type **0.5** and press <Enter> and the required reduction will appear.

Take a second copy of the blade on another part of the drawing sheet and use the ARRAY feature as before to draw a twin propeller, then a three-bladed propeller (Fig. 2.13). Note that if you draw a component half size, then all aspects of the construction are reduced in

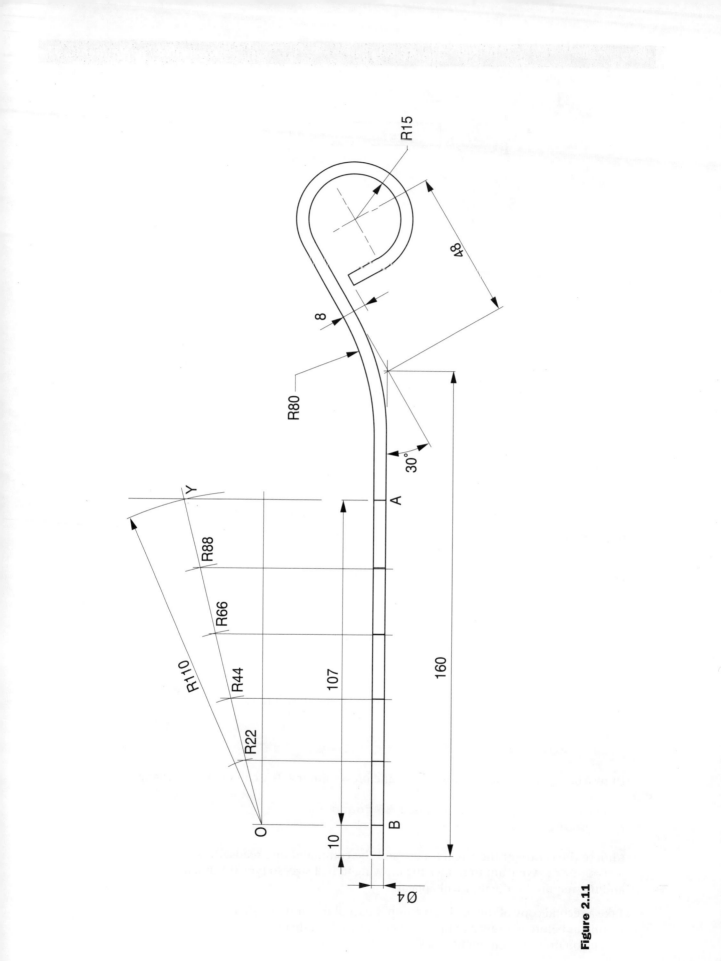

Figure 2.11

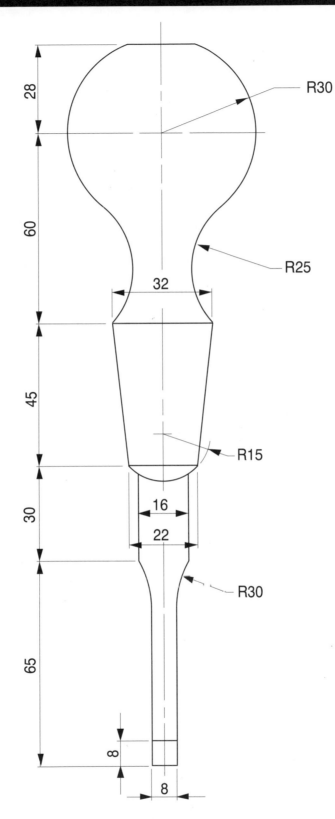

Figure 2.12

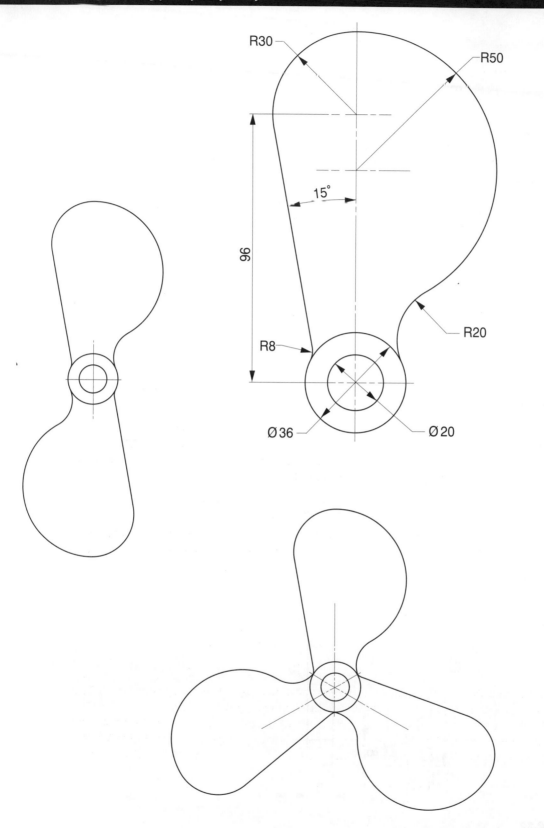

Figure 2.13

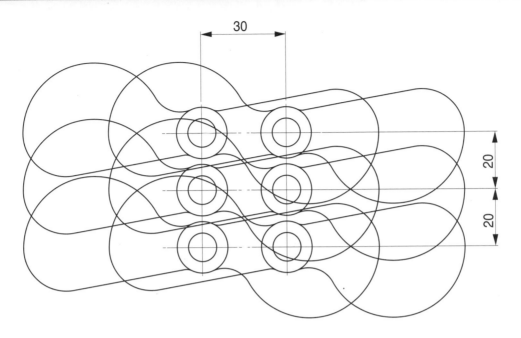

Figure 2.14

the same proportion. Thus if you require comparable linework it will be necessary to again use the EDIT POLYLINE feature to increase the usual line thickness to 0.4 mm.

At half size, the area covered by the drawing will also be one-quarter of its original area, so that experimenting with scale values may be necessary. Some people have difficulty in looking at a half-size drawing and then comprehending exactly how big the component actually is.

In addition to the polar array previously drawn, we can also use a RECTANGULAR ARRAY function. Copy the twin propeller onto another file and turn it through 90°. As you can see from Fig. 2.14, we require two columns 20 mm apart and three rows 20 mm above each other. Select **Array** from the **Construct** menu and read at the command line

```
Command: _array
Select objects
```

Draw a box around the propeller and click on the left mouse button:

```
Select objects: Other corner: (Quotes a number) found
Select objects
```

Click again to enter the information and command line reads `Rectangular or Polar array (R/P) <R>:`. Type **R**. The command line replies `Number of rows (---) <1>:`. Enter **3**. The command line replies `Number of columns (111) <1>:`. Enter **2**. The command line replies `Unit cell or distance between rows (---):`. Enter **20**. The command line replies `Distance between columns (111):`. Enter **30** and the arrangement is drawn.

Ellipse

One of the features of CAD which is valuable to the draughtsman is its ability to draw ellipses. An ellipse has two axes, known as the major and minor axes, and two focal points.

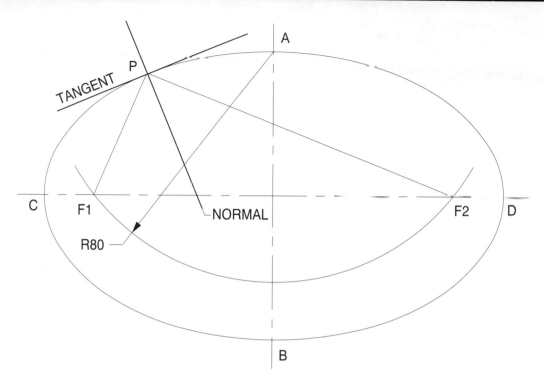

Figure 2.15

An ellipse can be drawn if you are given the dimensions of the two axes. The focal points are found by drawing an arc from one end of the minor axis which cuts the major axis at two points F1 and F2 (Fig. 2.15).

Exercise

Draw an ellipse with major axis CD = 160 mm and minor axis AB = 100 mm. Select **Ellipse** from the **Draw** menu. The command line reads

```
Command: _ellipse
<Axis endpoint1>/centre:
```

Click on a suitable position for point C and command line reads Axis endpoint 2:. Click on point D 180 mm to the right and command line changes to <Other axis distance>/Rotation:. Click on point A and the ellipse is drawn.

On Fig. 2.15, P is any point on the ellipse. The sum of lengths F1–P + P–F2 is equal to the length of the major axis CD. The normal to the point P can be found by bisecting the angle F1–P–F2 and the tangent constructed by drawing a perpendicular to the normal at point P. Add both of these important lines to your drawing.

A typical example could be as follows. Fig. 2.16 shows a special spring made from steel wire. Its shape consists of a true semi-ellipse ABC, of minor axis AC, which is in tangential contact with the true part-ellipse CDE. AC is the normal to CDE at C. F1 and F2 are the focal points of the part-ellipse CDE. Draw the spring to full size.

Note that it is necessary to establish the position of points D G and E. Measure and add the distances F1–C and C–F2 together, divide them by two and describe an arc from either focal point to give point D. Using the same radius along the horizontal centre line and from the centre point O, determine the positions of points G and E.

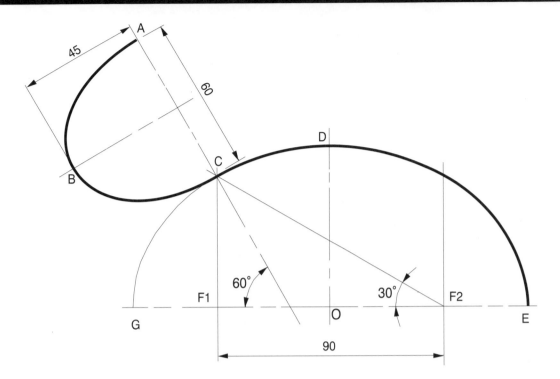

Figure 2.16

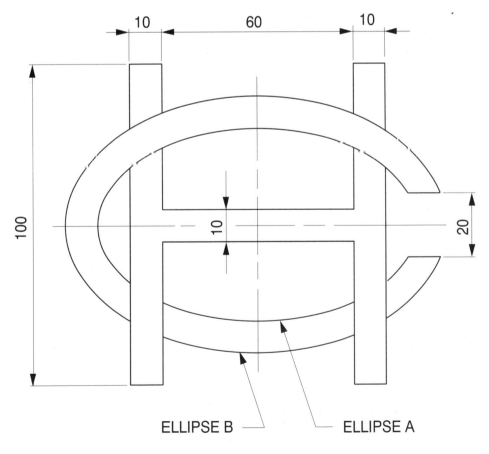

ELLIPSE B ——————   ————— ELLIPSE A

Figure 2.17

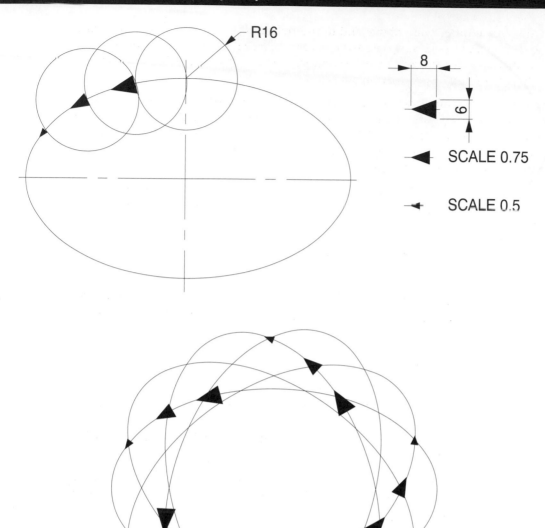

Figure 2.18

Figure 2.17 shows a logo for a Hotel Continental and consists of two ellipses. Ellipse A has a major axis of 100 and minor axis of 60. Ellipse B has a major axis of 120 and a minor axis of 80. Draw the profile of the logo and fully dimension the solution. Use the BREAK and TRIM features to delete the contours where the letter 'H' is positioned.

The example in Fig. 2.18 shows a sign designed from arrows in elliptical orbit. The construction steps are as follows. Draw an ellipse with major axis 100 mm and minor axis of 60 mm. Add three circles of 16 mm radius with their centres on the ellipse. Draw one arrow 8 mm long and 6 mm wide.

Choose **Copy** in the **Construct** menu and the command line reads Select objects:. Draw a box to enclose the arrow by clicking on a point below the bottom left-hand corner of the arrow and another point above and to the right of the arrow. The command line now reads <Base point or displacement>/Multiple:. Click on the arrow tip. The command line reads Second point of displacement:.Click on a point about 15 mm vertically below and the copy arrow appears.

It is now necessary to reduce the size of the arrow so select **Scale** from the **Modify** menu and the command line reads Select objects:. Draw a box as described above to enclose the arrow and the command line reads Base point:. Click on the tip of the arrow and the command line reads <Scale factor>/Reference:. Type 0.75 and press <Enter>. The reduced size arrow is drawn.

Repeat the procedure for the third arrow with a scale factor of 0.5. Select the **Move** command from the **Modify** menu to reposition the arrows on the ellipse as follows. The command line reads Select objects:. Draw a box enclosing the arrow and press <Enter>. The command line reads Base point or displacement:. Click on the centre point of the tail of the arrow and the command line changes to Second point of displacement:. Reposition the arrow on the ellipse.

The arrow will now be in the horizontal position with its tail on the correct spot but needs to be rotated so that its tip also lines up with the profile of the ellipse. Select **Rotate** from the **Modify** menu and the command line reads

```
Command: _rotate
Select objects:
```

Draw a box enclosing the arrow and enter and command line reads Base point:. Click on the centre of the tail and the command line reads <Rotation angle>/Reference:. The arrow will now rotate so reposition the tip on the ellipse profile and click again to enter.

Repeat the procedure for the other arrows, and erase the three circles used for marking the points. The ARRAY command previously described can now be used to draw five

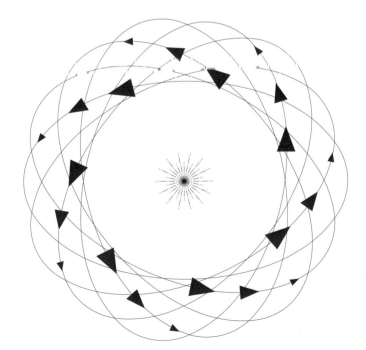

Figure 2.19

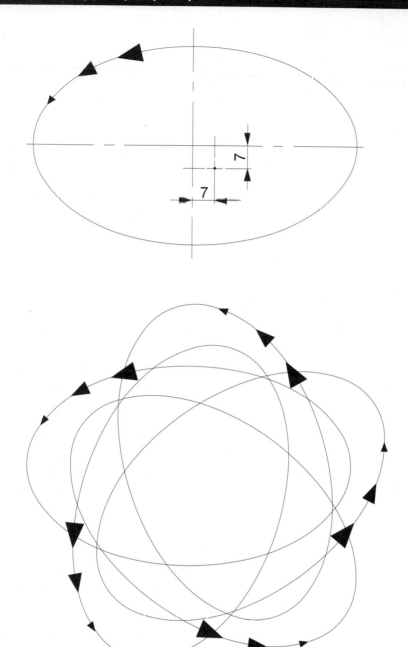

Figure 2.20

assemblies around the intersection of the major and minor ellipse axes. Clean up unwanted details to leave the drawing as illustrated.

Fig. 2.19 shows another application with seven arrays. A rosette pattern is formed by leaving parts of the original centre lines in place.

The ARRAY feature can be arranged about other points and in Fig. 2.20 the centre point has been moved 7 mm to the right and 7 mm below the original.

Wire parts

Many components are formed from coils of wire using high-production techniques. Basically, wire is automatically fed from a coil through sets of straightening rollers and released into a wire-forming machine and bent beyond its elastic limit into the required shape. Special-purpose machines incorporate cam-operated press tools, cut-off devices and means of ejecting the completed components.

When metal is permanently bent into a new shape, one side of the bend is subjected to compressive forces and the other side to tensile forces. Between the sides will be a neutral axis and its position varies with different materials. The production engineer will study a design drawing and make a bending allowance when determining the length of material required. The length of the bent component will not necessarily be equal to the length of the centre line.

Impossible dimensional restrictions should not be introduced, but good use of the tool possibilities will ensure the lowest costs. The manufacture of springs is a typical example where specialist manufacturers advice should be obtained.

Finished components may also be flattened with eyes pierced through the material, as for example on a needle. Swaging deforms metal by using compressive forces, e.g. in wire nail heads.

A large paper clip (item 1) and a wire form (item 2) with two eyes are shown in Fig. 2.21, and these have been bent from 1 mm wire. The dimensions refer to the centre lines of the components. Set out the profiles and use the EDIT POLYLINE feature in the Modify menu to finish your drawings.

The other components are formed from 4 mm wire. You can draw these components in several ways using the FILLET feature in item 3 and the FILLET and MIRROR features in item 4. In the case of item 5, the previous example can be modified by ERASURE and then replacement using the COPY command.

Layout exercise

Fig. 2.22 shows the dimensions of a wire hanger and four steps in the construction of the drawing. It is often necessary to prepare instructional drawings on a step by step basis, and CAD with its copying facilities is a very useful tool.

1. Draw the layout as shown in step 1. Use the TANGENT feature for the lines and add the normals from the circle centres.
2. Make a separate copy of your drawing and draw the polylines.
3. Copy the second drawing, erase the unnecessary detail and line in the arcs.
4. Copy the third drawing and repeat the process. Take a copy of this drawing and use the Mirror FEATURE to complete the given original drawing.

You will now have five small drawings somewhere on the screen and the final part of the exercise is to rearrange them using the MOVE feature to position them symmetrically on the A4 sheet. The **Move** feature is selected from the **Modify** menu, and at the command line you will read

```
Command: _move
Select objects:
```

In a similar manner to the operation for using the **Copy** feature, just draw a box to enclose the item to move, click with the left-hand mouse button at opposite corners of the box, press <Enter> and the command line reads

```
Select objects: Other corner: 30 found
Select objects:
Base point or displacement:
```

For example, click on the centre point of the R10 radius and the command line reads

```
Base point or displacement: second point of displacement:
```

Move the construction to your required new position, click and press <Enter>. If you are now uncertain quite where this will finally be, remember that you can always change it again an unlimited number of times, but choose one of the main lines on the grid. Use your grid to position the other views with the dimensioned drawing centrally at the top, as shown in the given figure.

These manipulations are a necessary part of the preparation of drawings and illustrations for use in desktop publishing applications.

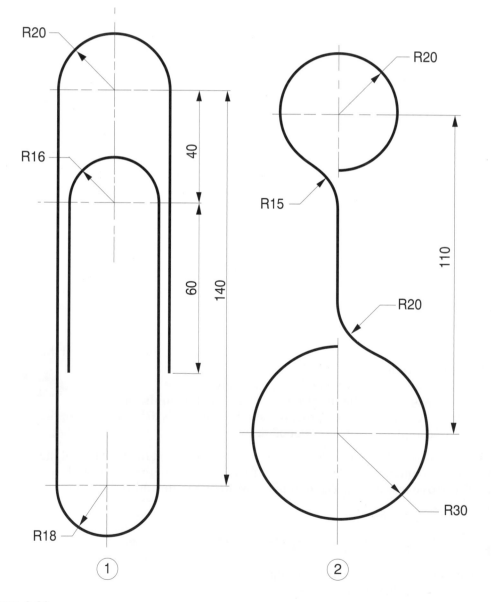

Figure 2.21

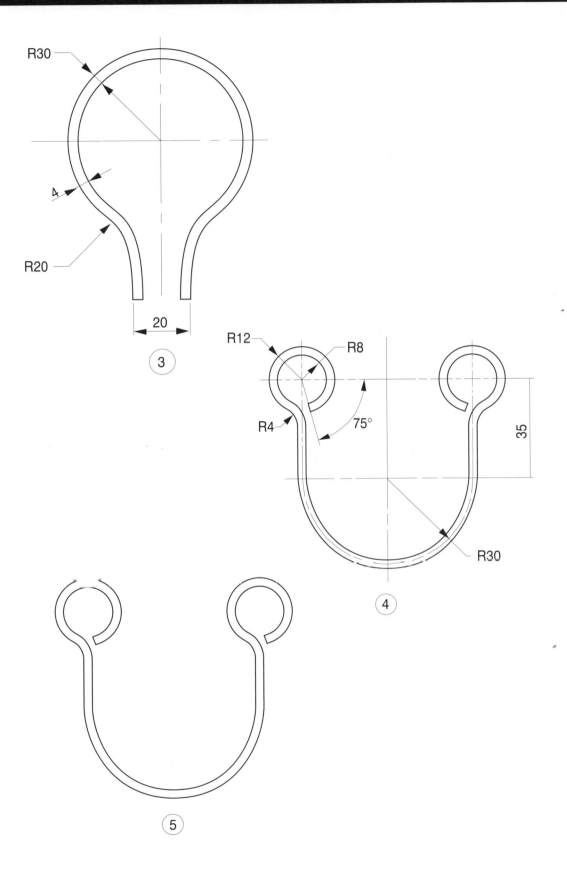

Figure 2.21 (continued)

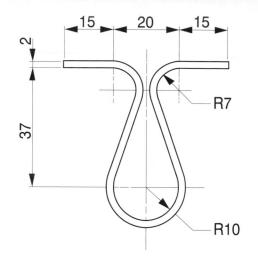

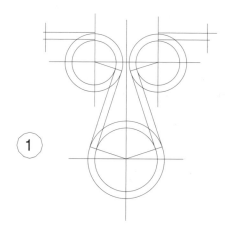

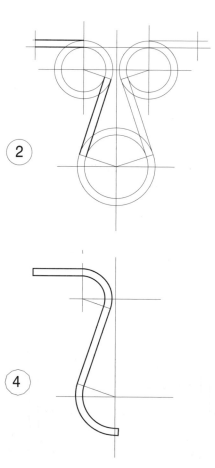

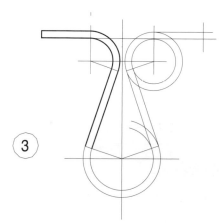

Figure 2.22

DOUBLE LINE command

Figure 2.23 shows the dimensions of a spring to be formed from 4mm wire. This component introduces the DOUBLE LINE command and for this first example we will restrict the application to the straight lines at each end of the spring.

Set out construction lines to fix the centre, the position of the angle and the ends of the spring as shown in stage 1. Draw circles of R28 and R32 and break them at the bottom right-hand corner since we are going to use the FILLET command at each side. In the **Draw** menu select the **Double Line** option. The command line reads

```
Command: _dline
Break/Caps/Dragline/Offset/Snap/Undo/Width/<start point>:
```

We need first to specify a width so type **W** and press <Enter>. Note that the line reads

```
Break/Caps/Dragline/Offset/Snap/Undo/Width/<start point>: W
New DLINE width<8>:
```

The previous line width was 8, so now type **4** and press <Enter>. Note change to

```
Break/Caps/Dragline/Offset/Snap/Undo/Width/<start point>:
```

Click on the point at the bottom of the spring in the 6 o'clock position and the next command reads

```
Arc/Break/CAps/CLose/Dragline/Snap/Undo/Width/<next point>:
```

Click with the left button about 35mm above as shown. Click again with the right-hand button to enter and the command line returns to `Command:`.

We can re-enter the previous command again by clicking with the right mouse button and repeating the process. If the SNAP button was engaged you will need to turn it off to ensure that the angle of 30° is exactly positioned on the vertical line through the end position on the component. If this has been done, then the last click during the process will close off the end of the parallel lines.

Use the FILLET command to insert the two pairs of radii. This needs to be done using radii of 18 and 22 since the mean radius of the centre line is 20mm. Use the EDIT POLYLINE command to give a finished drawing.

Double arc construction

Double lines can be arcs or straight lines and Fig. 2.24 shows a scale drawing of a layout for a model train set to illustrate some of the features available.

Draw a 100mm diameter circle through points A, B, C and D. Select **Double Line** in the **Draw** menu and the following options will appear at the command line:

```
Break/Caps/Dragline/Offset/Snap/Undo/Width/<start point>:
```

Click on point **A** and the command line reads

```
Arc/Break/CAps/CLose/Dragline/Snap/Undo/Width/<next point>:
```

Type **A** and press <Enter>, then **W** and <Enter> and command line reads `New DLine Width:`. Type **4**.

Double lines can be capped or left open so type **CA** and the command line gives the options `Both/End/None/Start/<auto>:`. Select None and type **N**. Click on point B, followed by C for the end point and a double semicircle is drawn. Type **CL** and the circles are completed.

Copy the three circles 110mm to the right to give E, F, G and H. Construct the three tangent lines separately using the OBJECT SNAP feature. Insert the double lines above for the train sidings. The three junctions can be completed using the FILLET feature with radii of 52, 54 and 56mm.

When using the FILLET feature you will find that one line is deleted which is required for the next step. This can easily be replaced using the EXTEND option. Line in the outlines with polylines of thickness 0.4 mm as shown in Fig. 2.25.

Gear wheel assembly

Fig. 2.26 shows the assembly of two gear wheels in mesh and notes associated with this branch of technology. Fig. 2.27 gives dimensions of the shaft details and tooth profile.

Draw the assembly using the following procedure.

1. Gear teeth are of involute form. Imagine a taut piece of string wound round a cylinder. As the end of the string unwinds, it follows the path of an involute. The diameter of this cylinder in gear wheel terminology is known as the base circle diameter, and the involute form is generated from this circle. In the example the 35 mm radius is an approximate dimension for a wheel of this size. Set out the profile by fixing point B, then point C. From C, use a radius of 43 mm to cut the radial line from centre O and fix the centre of the root radius. Determine the tangency point and blend the two radii to give the side of the tooth. You can of course draw one side and then use the MIRROR feature for the other side. Remember to use the SNAP feature whenever possible and ZOOM to keep a check on accuracy. Clean up any unwanted construction lines.
2. Use the ARRAY feature to draw eight teeth round centre O.
3. Make a copy of the wheel on the right-hand side.

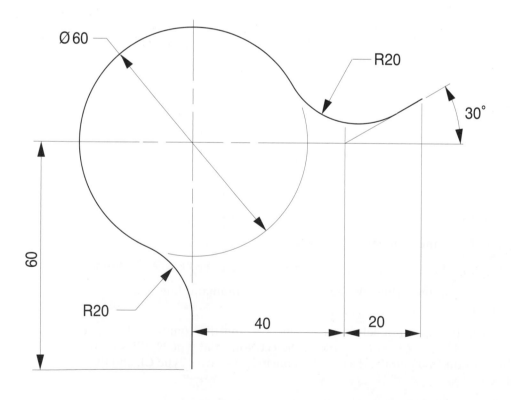

STAGE 1

Figure 2.23

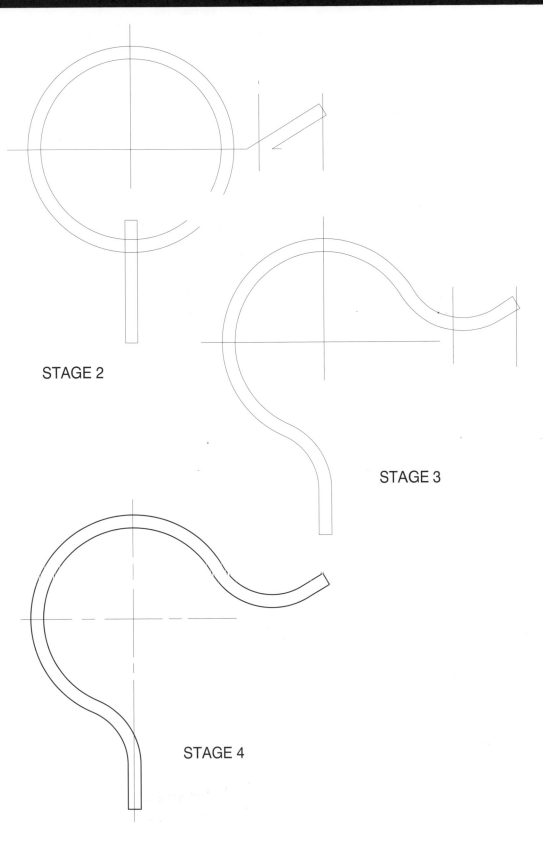

STAGE 2

STAGE 3

STAGE 4

Figure 2.23 (continued)

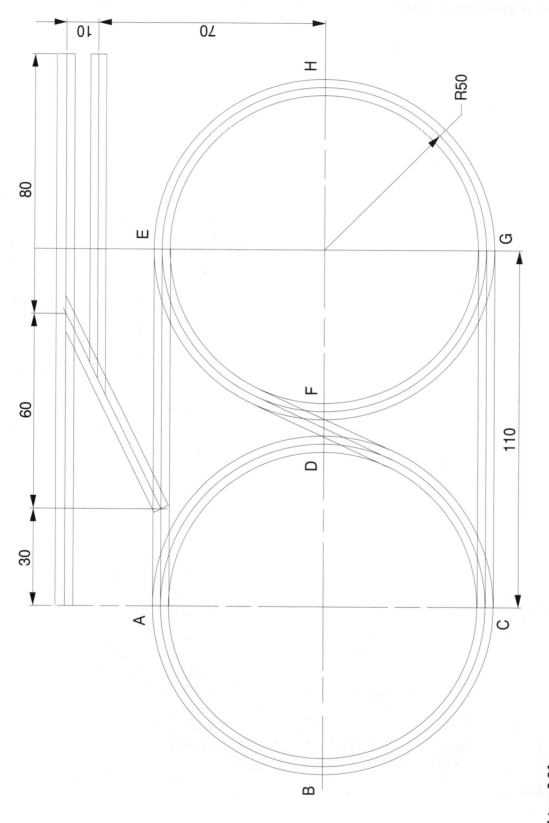

Figure 2.24

Figure 2.25

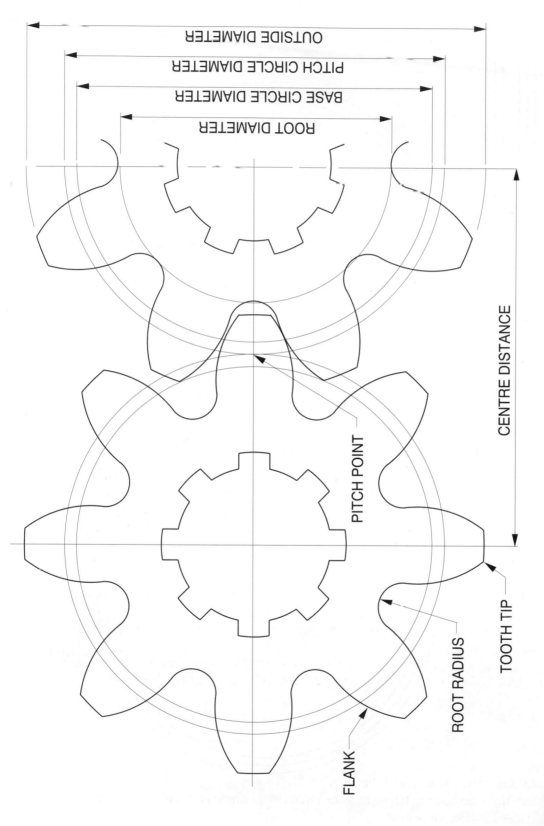

OUTSIDE DIAMETER

PITCH CIRCLE DIAMETER

BASE CIRCLE DIAMETER

ROOT DIAMETER

CENTRE DISTANCE

PITCH POINT

FLANK

ROOT RADIUS

TOOTH TIP

Figure 2.26

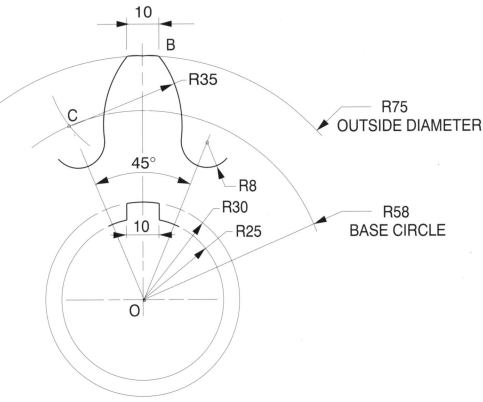

Figure 2.27

4. Gear teeth are positioned round a circle known as the pitch circle. Measured round the pitch circle circumference, the width of a gear tooth is equal to the width of the space between teeth. Gear wheels and their teeth of course vary in size. The pitch circle diameter divided by the number of teeth is known as the 'module' and wheels will mesh together if their modules are the same. If two standard involute gear wheels of different sizes are in mesh, then the distance between their wheel centres is equal to the sum of their pitch circle radii.

5. In this particular example the pitch circle diameter is 136 mm. Draw pitch circles on both wheels.

6. When wheels are in mesh, the imaginary pitch circles touch at the pitch point and this is shown on the drawing.

7. Use the ROTATION feature to rotate the right-hand wheel and note that this can be set at exactly 22.5°. Choose **Rotate** in the **Modify** menu and the command line reads

```
Select objects:
Base point:
<Rotation angle>/Reference:
```

Type **22.5** and press <Enter> and the wheel will rotate as required.

8. Now move the wheel to the left so that the pitch circles touch at the pitch point. Delete the right-hand half of the wheel.

9. Add the dimensions and details of gear wheel terminology to your finished drawing.

10. The wheels are designed to fit and rotate on splined shafts.

UNDO, REDO BREAK, EXTEND and TRIM features

When engaged on constructional and design work it is necessary to modify drawings, and this involves editing linework. You will also find that when you change one item, there is invariably a knock-on effect and further changes are required. Modifications to existing drawings are a large part of the draughtsman's work. It will often be possible to take an existing drawing, delete a few lines, add a few lines, change a few notes and a satisfactory new drawing is the result.

The UNDO feature is a blessing. We all make mistakes, but fortunately the computer remembers each step we make, as we make them, and these can be methodically undone.

If the command line reads Command:, Click on **Undo** in the **Edit** menu and the command line reads

```
Command: _U
GROUP
Command:
```

Press <Enter> and the last drawing operation will disappear from the screen. The command line now reads

```
U GROUP
Command:
```

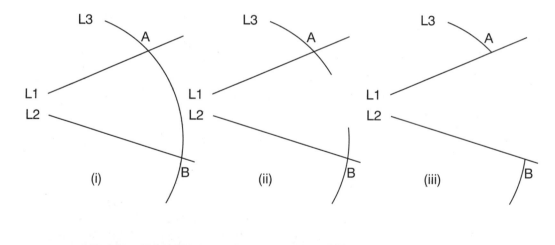

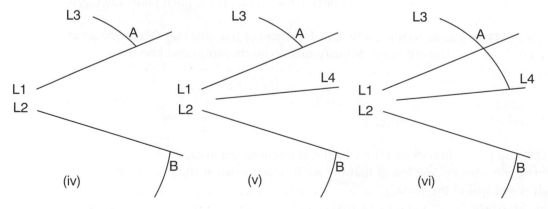

Figure 2.28

U GROUP means that the program has just removed the last line, but the expression will change according to your last drawing operation, and that may be a circle, erasure, break, etc. Just experiment by drawing some random lines, arcs, breaks and erasures on the screen and then cycle through the UNDO feature and clean them, one by one off the screen, noting the 'U' references.

REDO operates in the reverse order, so experiment again to understand the sequence.

1. Figure 2.28 shows two lines L1 and L2 drawn at random and an arc which crosses both at any two points A and B. I wish to break the arc and trim the two parts back to the lines L1 and L2.
2. Choose the BREAK command and click on the arc at any two points between A and B. A gap appears and there are two short lines to trim back to L1 and L2.
3. Choose the **Trim** command in the **Modify** menu. Click the two lines you want to remain, namely L1 and the arc above line L1. Press <Enter> then click on the small line you wish to delete. Repeat for the other side.
4. Alternatively, choose the TRIM command and click on lines L1 and L2. Press <Enter> and click once on the arc between the two lines which is then deleted cleanly.
5. Now assume that a third line L4 is inserted between L1 and L2 and you wish arc L3 to be extended from above to meet it.
6. Choose **Extend** in the **Modify** menu. Click on L4 and the required arc L3 above. Press <Enter> and then click on the arc L3 which will then be extended as required.

Loci and mechanisms

A locus is a curve traced out by a point, line or surface moving according to a mathematically defined condition. Figure 2.29 shows a simple case where a link connected to a rotating wheel at end X is designed to slide through a swivel fitting at point P. The engineer requires to know the path traced out by point Y at the end of the link. The reason for this interest may be that a protective safety guard is needed to cover the machinery, or that this is the first part of a mechanical system of levers and all aspects of the movement must be known.

Set out the diagram to the dimensions shown and divide the circle into 12 equal parts. The length of the link is 270 mm. Draw a line from the 8 o'clock position through point P and mark the position of point Y. I find that it is easiest just to draw a circle of 270 mm radius and use the BREAK command to delete the unwanted part of the circumference. The radii will be set to repeat the procedure. Marking the intersection with a small 'donut' of inside diameter 0.5 mm and outside diameter 1.0 mm is sufficient. Repeat the procedure for the other points from 3 to 9 o'clock. Since the required curve will be symmetrical, the plotted points can be positioned on the other side of the centre line using the MIRROR command. Draw a polyline through the intersection points and modify this polyline with the EDIT POLYLINE option in the Modify menu to give a curve. Study the curve. Clearly, the most accurate curve will be plotted using more points. I have added just one between 4 and 5 o'clock. Experiment and note the results. It is also possible that one half of the curve will be better than the other. If this is the case, just delete the worst half and MIRROR the best to obtain two symmetrical parts.

Archimedian spiral

An Archimedian spiral is the locus of a point which moves around a centre at uniform angular velocity and at the same time moves towards or away from the centre. A typical example can be found in amusement arcade pin ball machines. In Fig. 2.30 a ball is

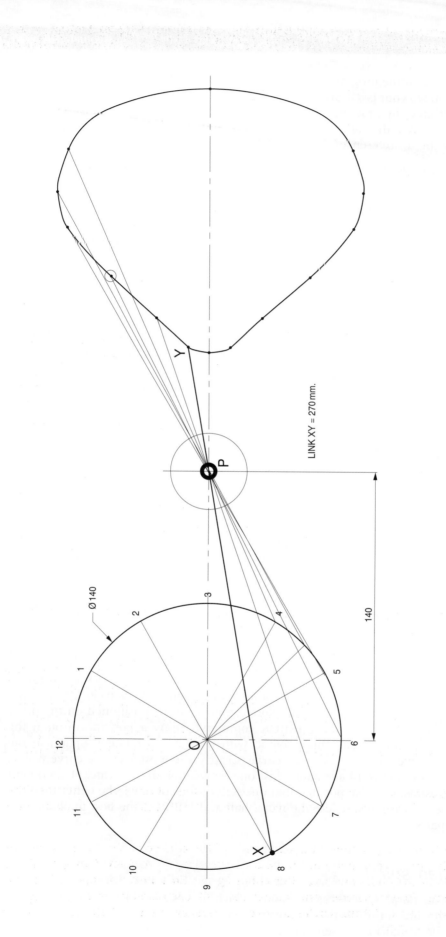

Ø140

LINK XY = 270 mm.

140

Figure 2.29

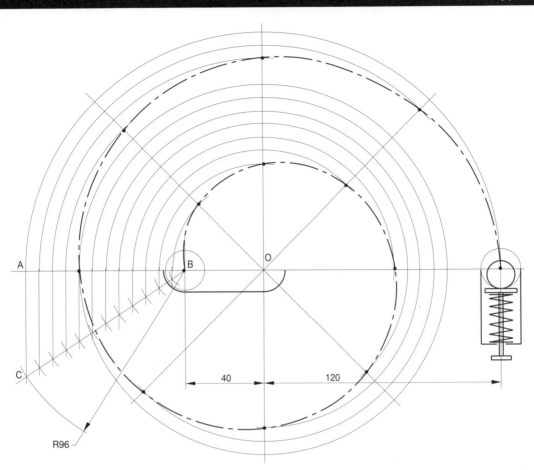

Figure 2.30

propelled around a 20 mm wide track for one and a half turns into a collecting dish and distributed into a network of pins and obstacles.

In relation to the centre, the ball moves a total radial distance shown by the length AB = 120 — 40 = 80 mm. The spiral can be set out using radial lines at 45°, then in one and a half revolutions the ball will move inwards in 12 steps. Set out the drawing and divide AB into 12 equal parts. From centre B draw an arc of 96 mm to intersect at point C with a vertical line from A, giving point C. Divide BC into 12 equal parts. From point B draw arcs of 8, 16, 24, 32, 40, 48, 56, 64, 72, 80 and 88 mm. (Note that in later versions of AutoCad LT a DIVIDE AND MEASURE feature has been included so that AB could be divided directly.)

Vertical lines from the intersections will divide AB as required. From centre O, draw arcs to intersect each of the 45° radial lines in turn. At each intersection draw a 20 mm circle.

From centre B draw a polyline connecting each of the circle centres. The polyline will be a succession of straight lines but can be converted into an Archimedian spiral if you type **F** for Fit at the command line after choosing the **Edit Polyline** option from the **Modify** menu. Draw polylines and repeat the edit command after connecting the circumferences of the 20 mm circles to give the outside profile of the track, shown separately on Fig. 2.31.

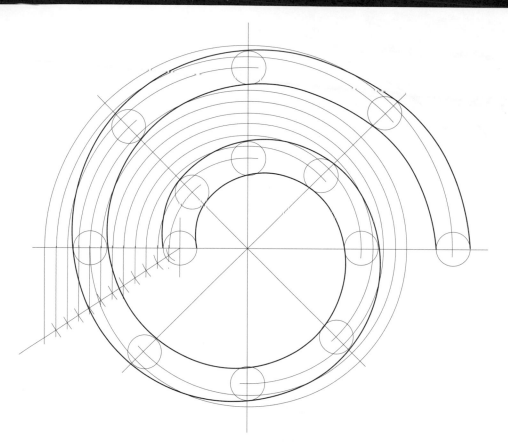

Figure 2.31

Helix

Figure 2.32 shows a curve generated on the surface of a cylinder by point P, which revolves with uniform angular velocity and at the same time moves axially with linear uniform velocity. The distance moved by the point in one revolution is known as the lead. If you take a cylinder and measure its diameter (D), the circumference length will be equal to $\pi \times D$.

Cut a triangular piece of paper as shown. Wrap it around the cylinder and the hypotenuse will trace the path of a right-hand helix. In this example the cylinder diameter and the lead measure 60 mm. Figure 2.33 shows a right-hand square-section helical chute of internal diameter 60 mm and external diameter 90 mm. This gives a square section with 15 mm sides. Divide the lead into 5 mm divisions.

Helical curves are generated from the four corners of the chute, but it is only necessary initially to draw two of them. Furthermore, it is necessary to plot only half of each helix since the MIRROR feature can be used to construct the symmetrical profiles. In the illustration there are one and a half turns and you will possibly find it advantageous to draw the helices side by side and use the COPY facility to build up the assembly in parts. Then use the BREAK command to remove hidden detail. Use the SNAP button where possible to maintain accuracy.

The illustration of the right-hand square-section helical spring in Fig. 2.34 can be constructed by taking a copy of the chute. Delete the lower third of this drawing and

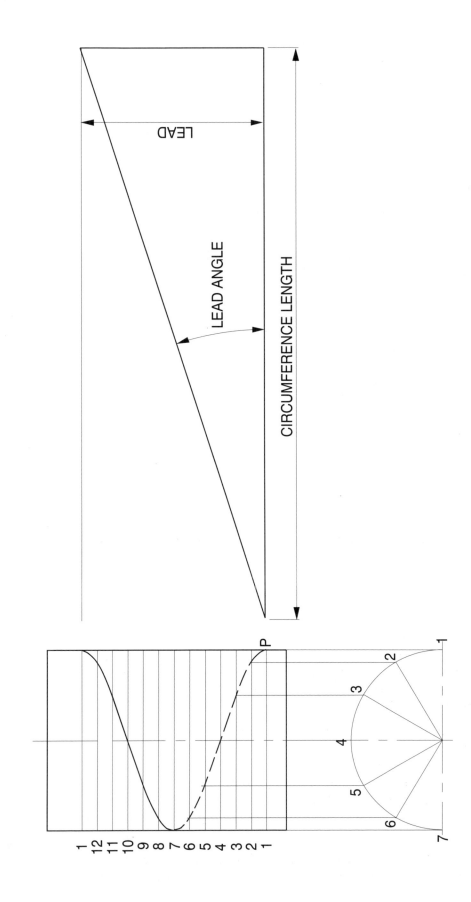

Figure 2.32

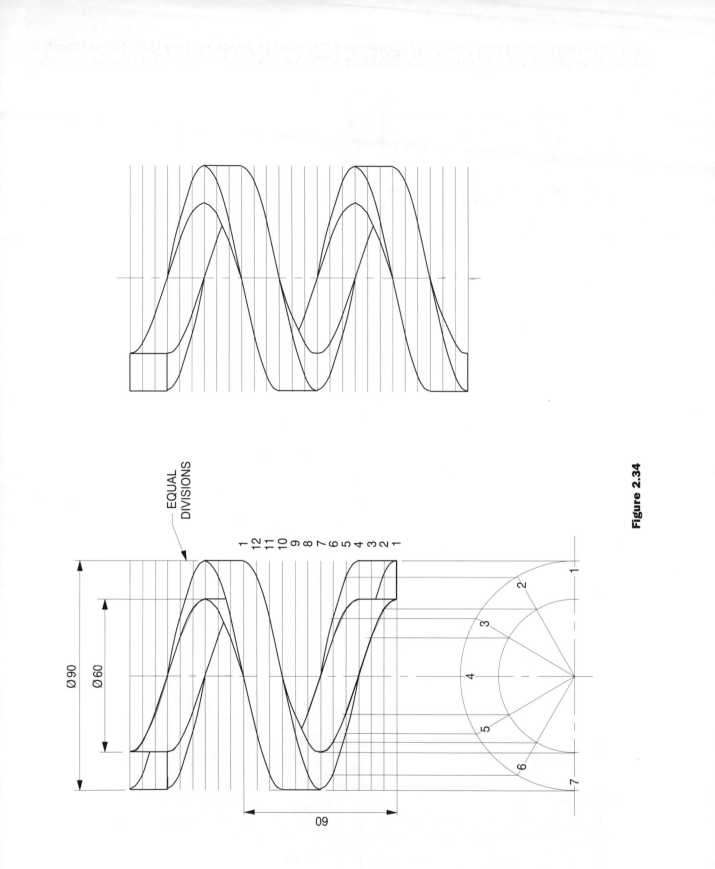

EQUAL
DIVISIONS

1
12
11
10
9
8
7
6
5
4
3
2
1

Ø90

Ø60

09

1
2
3
4
5
6
7

Figure 2.33

Figure 2.34

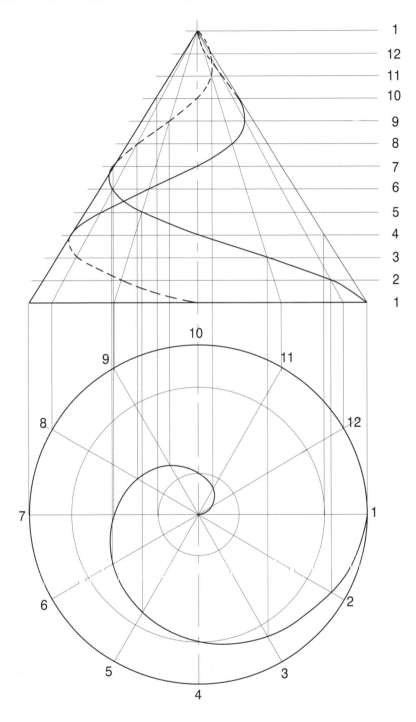

Figure 2.35

construct a copy of the remainder in the space above to give the two complete turns. Remove unwanted lines with the BREAK command.

Conical helix

The construction can also be applied on the surface of a cone and examples are given in Fig. 2.35. The base of the cone has a diameter of 120 mm and the height is 96 mm, which can be divided into 12 divisions of 8 mm width.

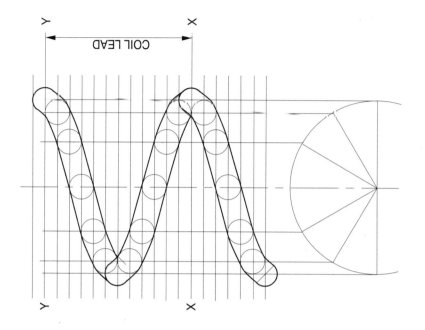

COIL LEAD

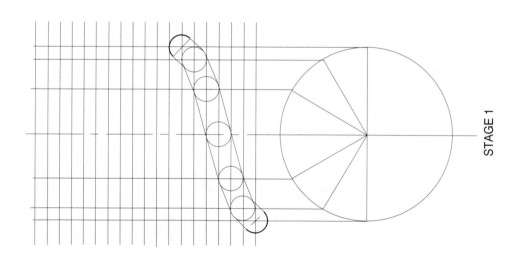

COIL LEAD

STAGE 1

Figure 2.36

The elevation shows two curves. A left-hand helix starts at point 1 and its projection is shown completely in the plan view. Note that at divisions 4 and 10 it is necessary to draw radii to establish the two points on the centreline. The other right-hand helix starts at point 10 and is constructed using the same method but with the direction of rotation reversed. Applications are found in fairground amusements and mechanical handling equipment.

Wire springs

A method of drawing wire springs is given in Fig. 2.36, where the points on the helix are projected as before and at each point a circle is drawn to represent the wire diameter. Tangents between adjacent pairs of circles need to be drawn to form continuous lines as shown in stage 1, which are then converted to polylines. Use the MIRROR and COPY features to add together the separate sections as mentioned previously.

Draw stage 1 for a spring with a mean diameter of 70 mm and a coil lead of 30 mm.

Chapter 3

Pictorial projections

This chapter introduces the following topics:

- Isometric principles and planes.
- Oblique, Cavalier and Cabinet projections.
- Planometric projections.
- COPY and ROTATE features applied to animation.

Isometric and oblique projections

Isometric and oblique projections are the most common methods of drawing and sketching, so that the three-dimensional proportions of a component can be seen in one illustration. These forms of projection do not take into account the element of perspective which a true artist produces. Their function is to convey understandable shape and form and instant recognition. The ability to read two-dimensional drawings and plans requires some study and appreciation of their construction in order to interpret the linework. If you hold a cube between your fingers and grip it across a diagonal on one of the faces, then gradually rotate it, you will be able to see the view in Fig. 3.1. You will appreciate that you are not looking at surfaces which are at right angles to your line of sight but three surfaces at an angle.

However, as you twist the cube you will see that a situation can be obtained where all corners of the cube appear to be the same length. They are not of course true lengths because the corners are angled away from you. In isometric projection we draw and measure along the three principal axes which are either vertical or 30° to the horizontal.

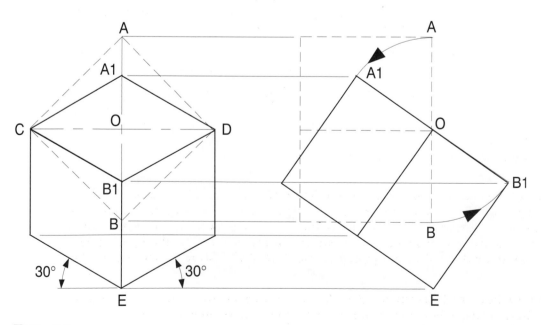

Figure 3.1

Drawing notes for Fig. 3.1

Construct this drawing as an exercise. Draw a cube of any convenient size with one face ACBD shown by the dotted line. Draw the dotted line AOB a suitable distance to the right having projected the corners A and B. Add the other dotted lines in the right-hand view to show the thickness and complete an end view of the cube. Imagine the cube to be rotated in the left-hand view across the diagonal COD until angle A1–C–O is equal to 30°. Complete this view with length A1–C equal to B1–E.

Project lines from points A1, B1 and E to the right. Lengths can be trimmed later. You can now physically rotate the cube in the right-hand view by the ROTATE command and using the point O as the centre of rotation.

Simply turn the cube anticlockwise until the lines intersect as shown. You have two views in first-angle orthographic projection which is described in a later chapter. The left-hand view in isolation is an isometric drawing of a cube. Line your work in using polylines.

Note that in your isometric drawing, the true length of the cube AC actually appears as C–A1, which is shorter in length. This fact is of academic interest only as we use the given true lengths in isometric constructions. AutoCAD can reduce and also enlarge drawings if necessary.

AutoCAD has some handy features to allow rapid draughting. The Drawing Aids box in the Settings menu provides a grid of lines at 30° to the horizontal and vertical lines, forming a grid of equilateral triangles so that lines can be positioned along the three axes for accurate measurement.

In the Draw menu we also have an ELLIPSE feature which allows the quick insertion of ellipses in the three planes. By using the <Ctrl> + **E** keys on the keyboard we can toggle and obtain the cursor lines for each of the three planes.

As an exercise to demonstrate basic Isometric drawing, draw a piece of tubing of 40 mm outside diameter, 30 mm inside diameter and 25 mm long in each of the planes (Fig. 3.2).

Open the **Drawing Aids** box from the **Settings** menu. Insert a SNAP value in the box which suits the required drawing dimensions and in this case 5 mm is appropriate. Insert a value of **10** in the **Grid** box. Click on the **Isometric Snap/Grid** box in the bottom right-hand corner and notice the 'X' value changes to 8.6603 in the SNAP box and to 17.3205 in the GRID box. Click **OK**.

The command line reads Command: `'_ddrmodes`. Use <Ctrl> + **E** and toggle until you read <Isoplane Left> at the command line and the cursor lines appear.

To draw the 40 mm ellipse, select **Ellipse** from the **Draw** menu and the command line reads

```
Command: _ellipse
<Axis endpoint 1>/Centre/Isocircle:
```

Select **Isocircle** and enter **I**. The command line reads Centre of circle:. Click on the chosen point and command line reads <Circle radius>/Diameter:. Enter **20** and the ellipse is drawn. Repeat the procedure and enter **15** for the internal 30 mm diameter. At a distance of 25 mm along the axis of the pipe and to the right, draw a second ellipse for the end of the pipe. Use the TANGENT feature to add the outside lines.

Choose **Line** from the **Draw** menu, then **Assist** and click on **Tangent** in the **Running Object Snap** dialogue box. Repeat for the other tangent. Remove unwanted linework and finish the drawing with POLYLINES.

Use the BREAK feature to insert a space in the unwanted part of the ellipse. (Stage 2 shows one end completed and the other end before trimming.) It is then necessary to clean up the two ends back to the tangency points. Use the TRIM facility in the Modify menu. Choose this option and the Command line reads

```
Command: _trim
Select cutting edge(s) . . .
Select objects:
```

Click on the two lines which meets at the tangency point and form the profile, then press <Enter>. The command line changes to

```
Select objects: 1 found
Select objects: 1 found.
```

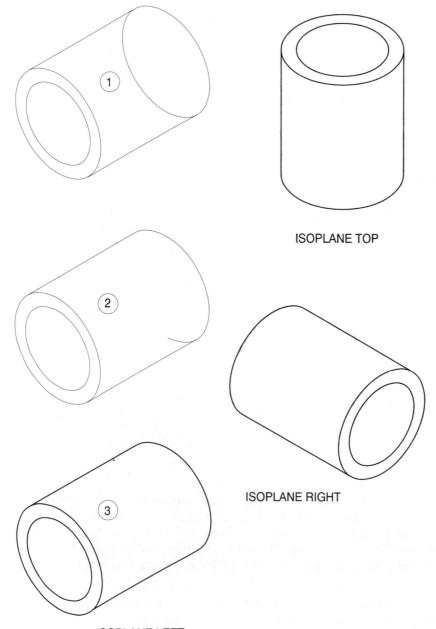

ISOPLANE TOP

ISOPLANE RIGHT

ISOPLANE LEFT

Figure 3.2

After clicking on the two lines to remain, press <Enter> and the following command appears: `<Select object to trim>/undo:`. Click on the small section of line to remove it. Use EDIT POLYLINE in the Modify menu to give the finished drawing.

Exercise 1: basic isometric drawing examples

- Figure 3.3 shows four solid components and each has been drawn on a 10 mm grid with a 5 mm snap value. Open the **Drawing Aids** box from the **Settings** menu and click the **Isometric/Snap Grid**.
- Ex. 1 can be drawn using the grid as a guide for the polylines.
- Ex. 2 requires you to choose ISOPLANE LEFT. Draw ellipses with 5 and 10 mm radii and delete the unwanted parts with the BREAK option.
- Ex. 3 uses ISOPLANE RIGHT with 5 and 10 mm ellipses.
- Ex. 4 uses both ISOPLANE RIGHT and ISOPLANE TOP. Complete one part and then toggle to the other plane.

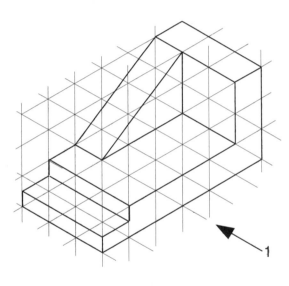

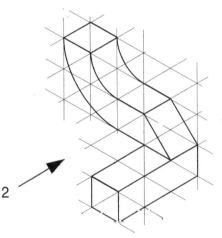

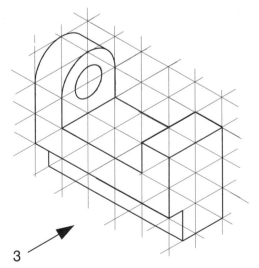

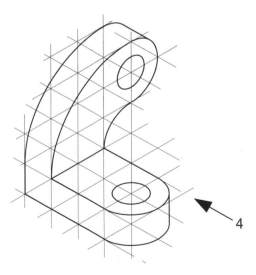

Figure 3.3

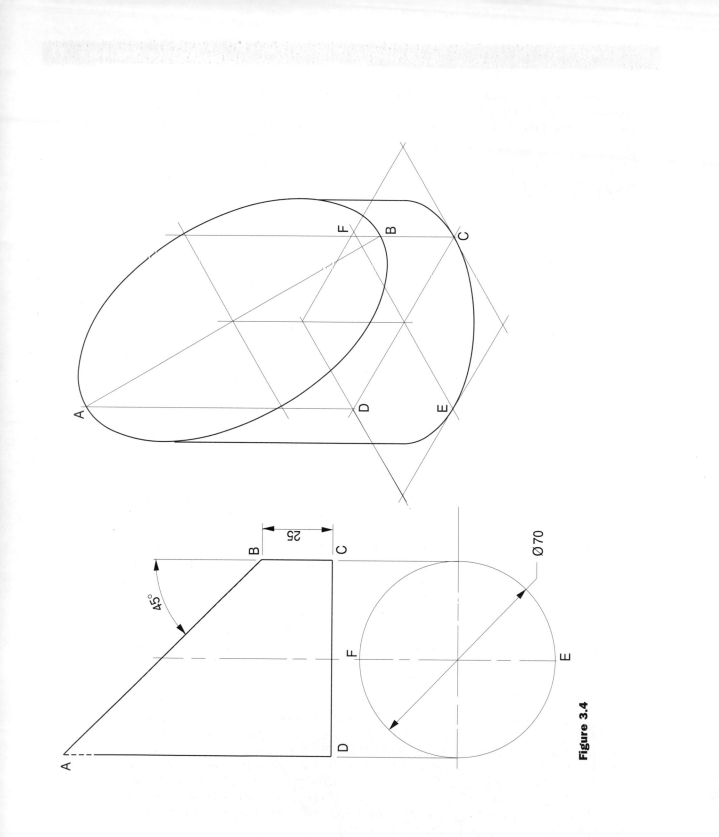

Figure 3.4

Ø70

45°

25

You will find it advantageous to always use the ZOOM feature to draw on the largest screen area possible. There is no point in struggling in a small area as errors become obvious to you when enlarged. Remember also to keep the SNAP button engaged if you can.

Figure 3.4 shows an isometric view of part of a circular bar. Copy the given view and prepare the pictorial base with lines at 30° to the horizontal. Distances DC = EF = 70 mm. DA in the pictorial view = DA = 95 mm in the orthographic view. Distances EF and AB in the pictorial view give the dimensions for the minor and major ellipse on the sloping face. It is not necessary to use the ISOMETRIC SNAP/GRID feature as lines at 30° can easily be drawn.

Both ellipses can be drawn using the **Ellipse** feature from the **Draw** menu. The Command line will read

```
Command:
ELLIPSE
<Axis endpoint 1>/Center:
```

Click the left mouse button on point B and the command line reads `Axis endpoint 2:`. Click on point A at the other end of the major axis and the command line changes to `<Other axis distance>/Rotation:`. Click on either end of the minor axis across the sloping face and the ellipse will be drawn.

Repeat for the base circle and use the BREAK feature to remove the unwanted part.

Oblique projection

In this form of pictorial projection, one face of an object is drawn in its normal front view and then its thickness is given by projection lines generally at 45°. This is probably the most common angle for convenience of drawing but others can be used. Figure 3.5 shows two illustrations of a cube. You will note that the right-hand cube in cavalier projection looks out of proportion if the lines on the oblique plane are drawn true to size. To improve the appearance, these true lines in cabinet projection on the left-hand view are halved in length. The problem arises due to lack of perspective in this style of drawing. To introduce perspective would complicate and considerably extend the draughtsman's work and make the job uneconomical. Cabinet projection is generally the accepted choice. If curves are present on one face it is often possible to present them in the foreground view but if they appear on two or three faces then they need to be plotted as shown in Fig. 3.5.

The example in Fig. 3.6 shows a cabinet projection of a cube and the method of drawing circles on each face. The cube has sides of length 120 mm with a circle of 100 mm diameter. The three corner lines of the cubes on the receding faces can be drawn as described before for a typed line input using **@60<45** at the command line. The dimensions for the distances between ordinates will be 10 mm measured along the 45° lines. Note that for the curve construction on the right-hand face, you will need to measure distances of 5, 10, 20, 30, 40, 50 and 55 mm. For convenience, I find it quick and easy to do this by drawing circles from point A, using the intersections with line AB to insert the ordinates, and then erasing the circles afterwards.

I have projected the construction lines from the front face giving sufficient points to construct the curves, and have lined in one of them. This curve is not an ellipse but can be constructed by drawing a polyline from point to point with a a succession of straight lines. Select the **Edit Polyline** feature in the **Modify** menu. The command line will read

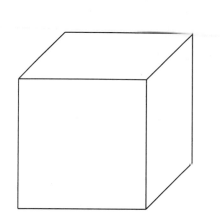

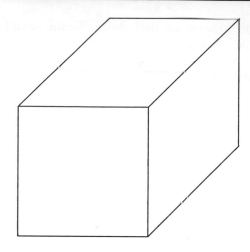

CABINET
PROJECTION

CAVALIER
PROJECTION

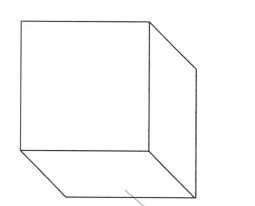

ALTERNATIVES

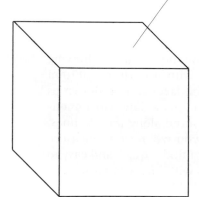

Figure 3.5

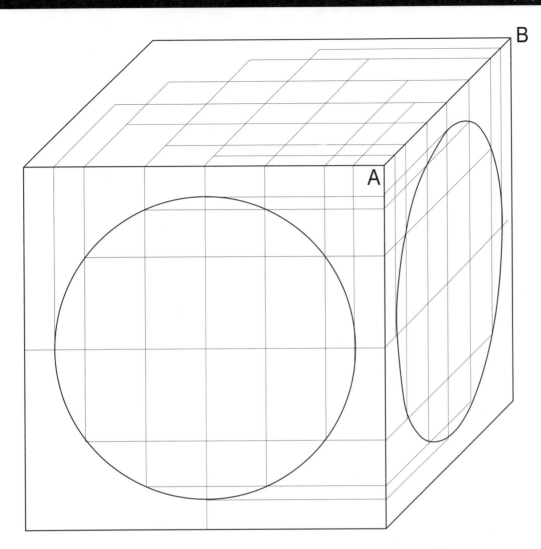

Figure 3.6

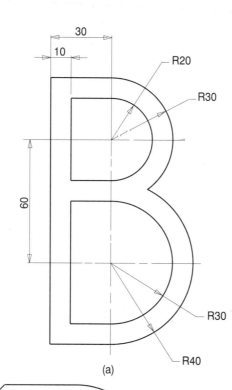

(a)

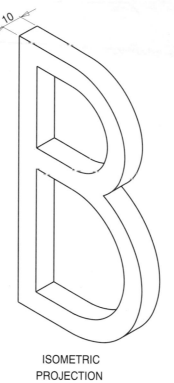

ISOMETRIC
PROJECTION

(b)

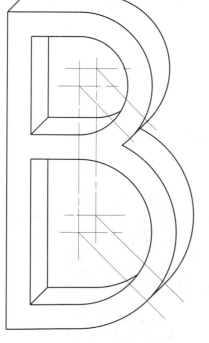

CAVALIER PROJECTION
(c)

CABINET PROJECTION
(d)

Figure 3.7

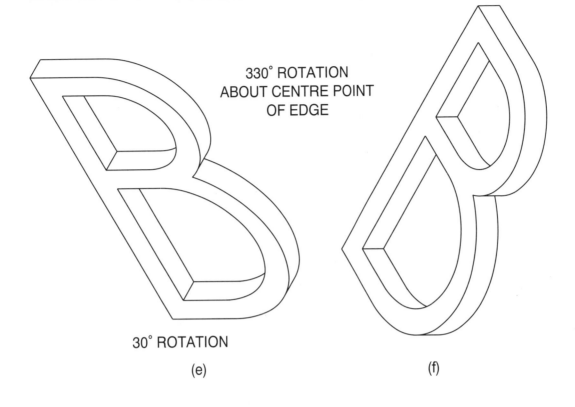

330° ROTATION
ABOUT CENTRE POINT
OF EDGE

30° ROTATION

(e)   (f)

Figure 3.7 (continued)

```
Command: _pedit Select polyline:
Open/Join/Width/Editvertex/Fit/Spline/Decurve/Ltypegen/Undo/
eXit<X>:
```

Type **F** for Fit and the straight lines will blend into a curve passing through the chosen points.

Exercise 2: draw the letter 'B' in cavalier and cabinet projection and compare the solutions with isometric projection

The profile and dimensions of the letter 'B' are shown in Fig. 3.7(*a*). This is an example of an 'extruded' section where the rear profile is the same as the front as for example in a tube or bar.

The cavalier solution requires a copy of the front view to be repeated 10 mm to the rear at 45°. Use the centre of the top semicircle as the datum and the COPY feature to draw a second profile. First ensure that the SNAP feature is engaged and click on the centre point. The 10 mm will now be measured from this established datum and along the hypotenuse of a 45° triangle, so the corners of the triangle will not all coincide with SNAP positions. Turn the SNAP feature off and for the second displacement type **@10<45** at the command line, press <Enter> and the second letter 'B' appears in its new position. Note the tangent lines to be added using the RUNNING OBJECT SNAP option between the pairs of construction lines. Finally edit the finished linework using BREAK and TRIM.

Make a second drawing for a cabinet solution with the reduced thickness. It is often the case that thicknesses are small, so use the biggest ZOOM possible for ease of

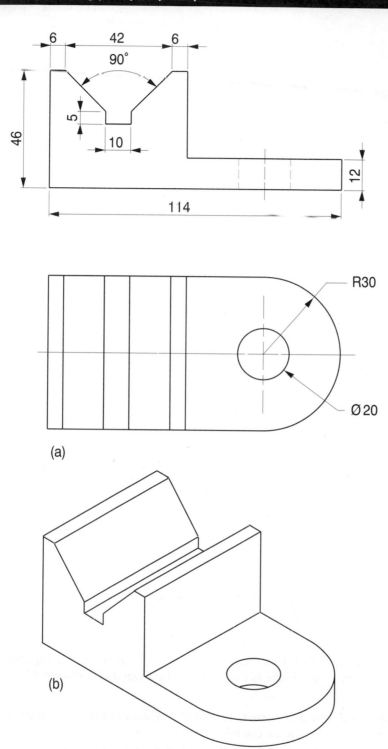

(a)

(b)

Figure 3.8

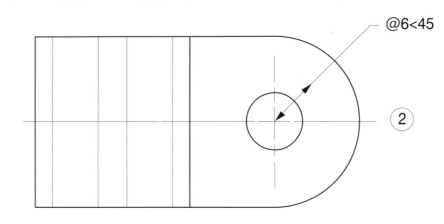

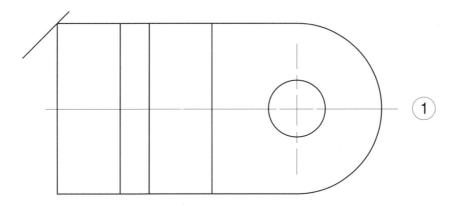

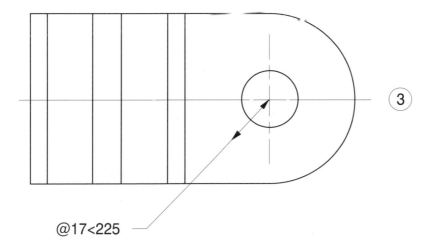

Figure 3.9

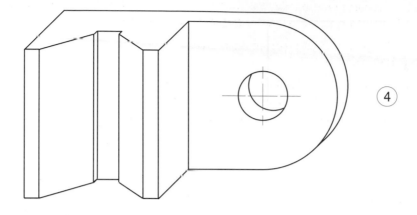

Figure 3.10

construction and accuracy. The isometric drawing requires the ellipses for the semi-circles to be drawn with the ISOPLANE LEFT feature engaged. Select a GRID value of 10 mm.

If you now compare the three views you will appreciate why the cabinet style is the one normally chosen with better proportions when an oblique view is required. In artistic work there is no reason why the ROTATE feature cannot be used to provide different orientations and effects. Experiment as shown in Fig. 3.7(*e*), (*f*).

Exercise 3: isometric and oblique projections of 'V' block

Draw the two views in orthographic projection of a 'V' block shown in Fig. 3.8(*a*). The isometric solution in Fig. 3.8(*b*) requires the ISOPLANE TOP feature to be engaged to construct the complete ellipse for the bolt hole and the semi ellipse for the outside profile. Copy these profiles 12 mm beneath and use the tangent feature for the 12 mm vertical line on the right-hand side. Use BREAK and TRIM to clean up unwanted lines. The 'V' profile can now be positioned using true dimensions along the isometric axes for the faces at 45°. In the oblique solution it will be advantageous to orient the illustration so that the bolt hole appears as a true circle.

Note that there are three distinct datum faces on the drawing, namely the top of the base, the bottom of the base and the top surface above the 'V'. In oblique projection the true distances between these datums will be halved. Make a copy of the plan view from the original drawing in the centre of a new sheet as shown in Figs. 3.9–3.10. Take another two copies and position them above and beneath as indicated since we can use parts of them and there is no need to redraw.

Use the centre of the bolt hole as the new datum. Select **Move** in the **Modify** menu and click on the hole centre with SNAP engaged. Turn SNAP off, then type **@6<45** at the command line and the outline will move to its new position. Repeat the procedure for the bottom view but type **@17<225** for the new position. Remember that the angles are measured anticlockwise from the horizontal.

In the centre view the top corner line has been drawn. Select **Draw** and **Polyline** for each in turn and type in the same values to establish the short lines shown. You will note that for convenience some unwanted lines have been deleted from the views.

Ensure that ORTHO is engaged and copy the bottom view so that its top left corner is positioned at the end of the angular line for the front face of the solution. You could of course use the MOVE command but I prefer not to do this since it is often easier to have two bites at the cherry if problems arise, since I still have the original.

Plot the profile of the 'V' on the top surface and transfer a copy to the bottom. Clean up the drawing with BREAK and TRIM.

Repeat the procedure by copying the top view in its new position and then add the short tangent to the two semicircles. Use the ZOOM feature to get the maximum size of image on the screen for clarity and engage SNAP for guaranteed accuracy whenever possible.

Further examples of isometric and oblique projections are shown in Figs 3.11–3.13.

In Fig. 3.14 a method of setting out the hexagon is shown by taking the distance across the flats and the distance across the corners from the orthographic view and marking out equal dimensions along the main isometric axes by the use of circles. To avoid confusing construction lines on the main drawing of many features, it is often easier to construct detail away from the main drawing and simply insert it using COPY or MOVE commands.

In the bracket shown in Fig. 3.15 there are two similar holes with centre lines, and the technique here is to draw one in a clear space away from the drawing and then use two copies in the correct position. In Fig. 3.16 the sloping face at 45° has the effect of producing a single line on the left side of the oblique view. An alternative solution can be presented with the receding side at 30° as shown in Fig. 3.16(*b*)–(*d*).

Drawing notes for Fig. 3.17

This exercise shows a small moulding to be drawn in isometric and oblique projection. Draw the given dimensioned example.

In each solution there are three vertical planes. For the oblique solution take a copy of the left-hand given view, to be used as a datum. Erase all dimensions. Position another copy 5 mm to the rear at 45° and use the BREAK command to remove the hidden parts. Add the top left and bottom right corner lines. Project forward the two semi-circles and reposition them at 18 mm towards the foreground. Then add the tapered side lines.

The isometric view needs to be set out using the isometric grid. You may find that it is easier to draw the ellipses away from the solution since only halves and quarters are needed. Use COPY and MOVE features to separate the required parts. The advantage of this method is to reduce the number of construction lines in such a small area. If construction lines have previously been drawn to establish the outline, then finished polylines can be repositioned with accuracy. Remember to use the SNAP feature with the appropriate settings and the ZOOM command.

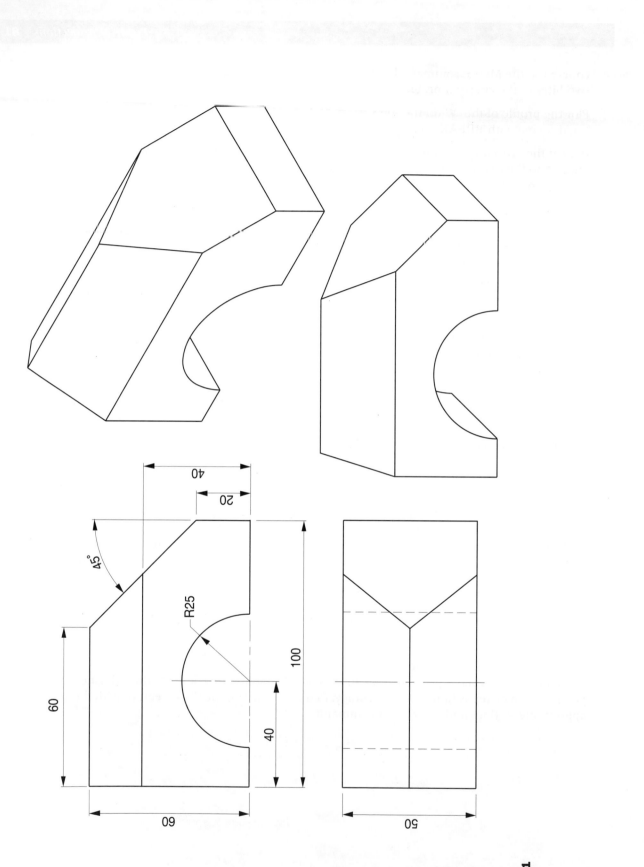

Figure 3.11

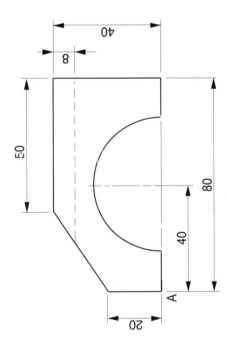

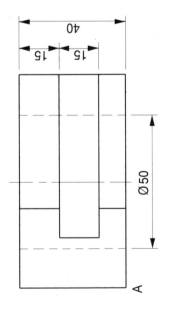

Figure 3.12

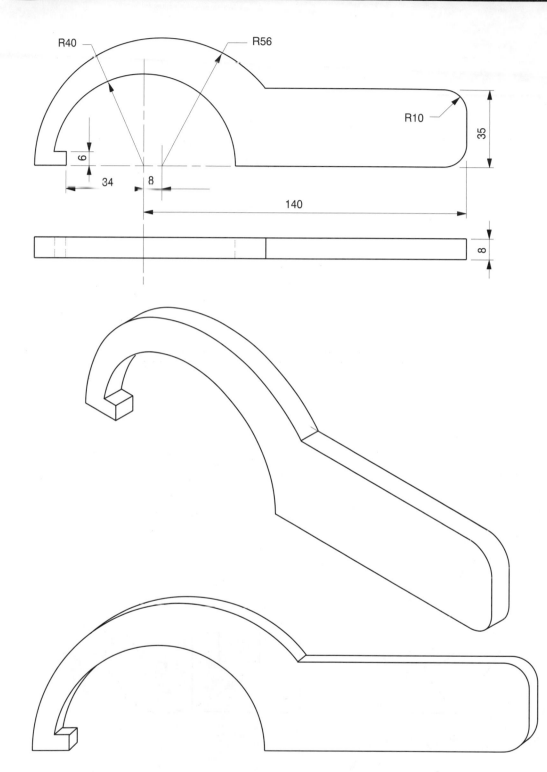

Figure 3.13

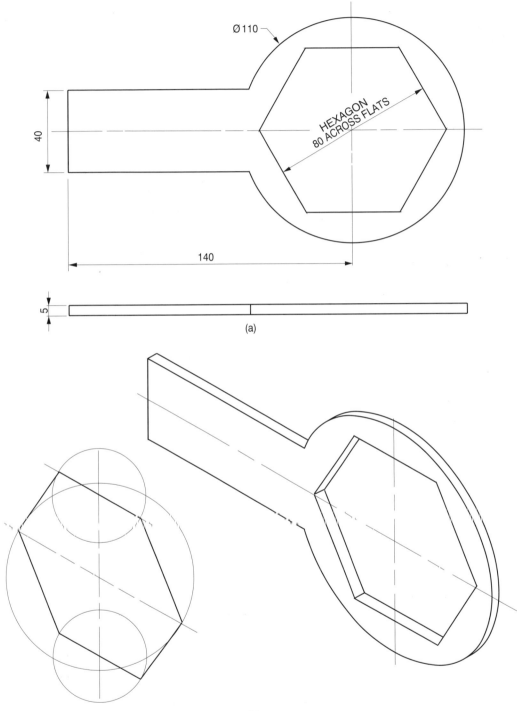

Ø110

40

140

5

HEXAGON
80 ACROSS FLATS

(a)

(b)

Figure 3.14

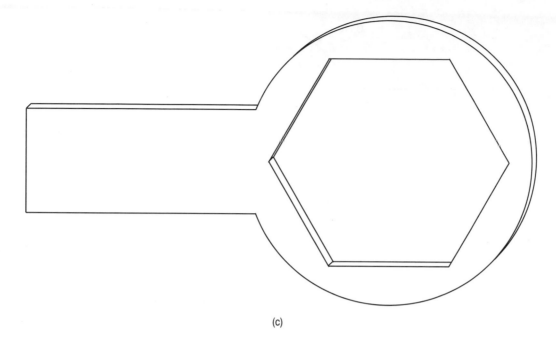

(c)

Figure 3.14   (continued)

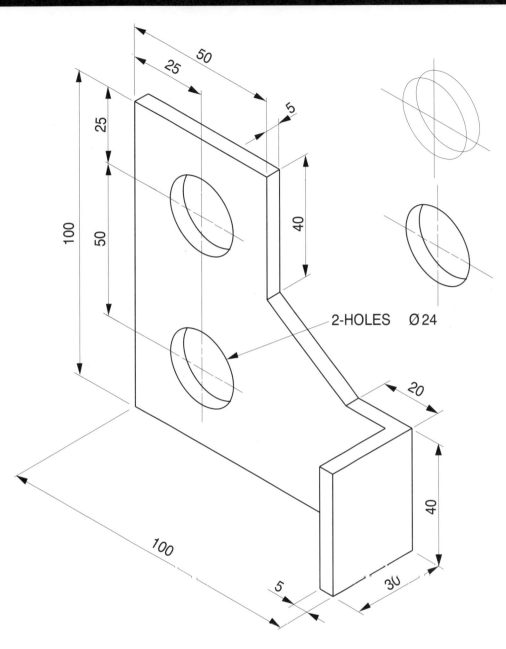

2-HOLES   Ø 24

Figure 3.15

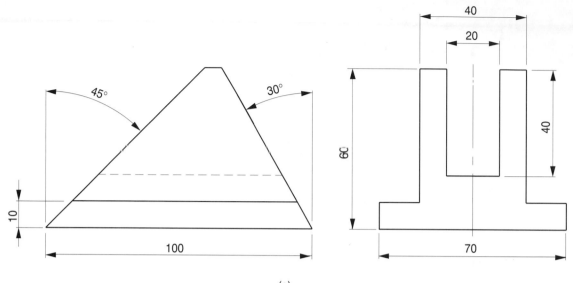

(a)

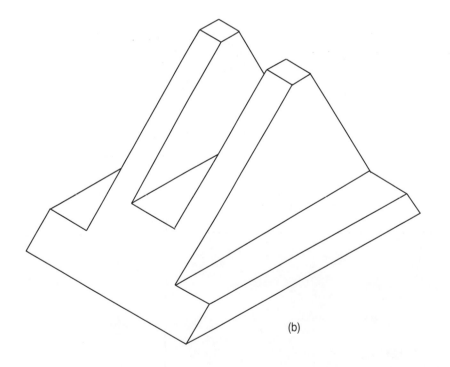

(b)

Figure 3.16

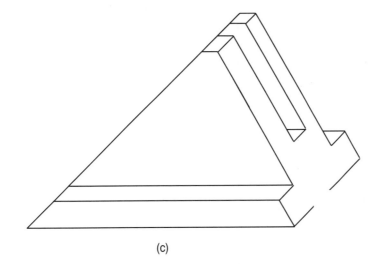

(c)

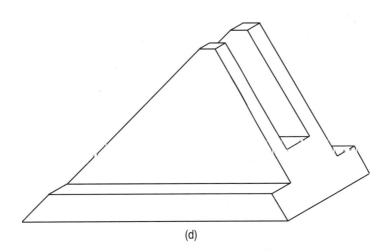

(d)

Figure 3.16 (continued)

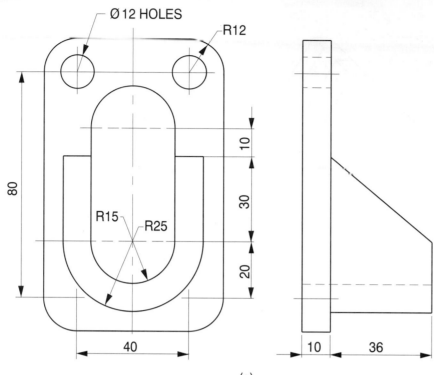

(a)

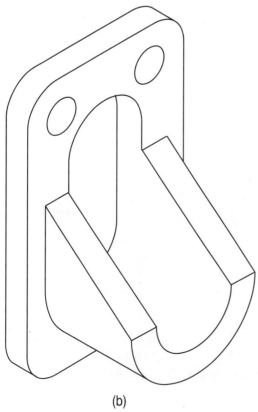

(b)

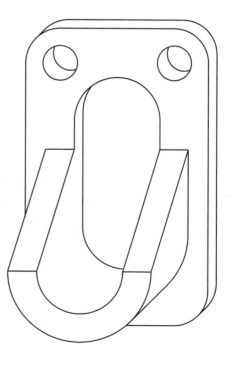

(c)

Figure 3.17

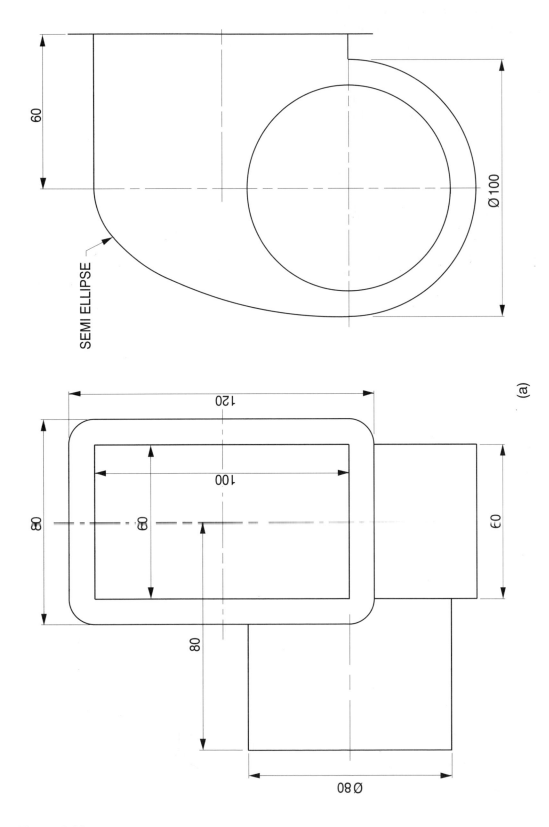

SEMI ELLIPSE

Ø 100

60

120

100

80

60

80

Ø 80

(a)

Figure 3.18

Motor housing

Fig. 3.18(*a*) shows the dimensions of a fan motor housing. An isometric presentation is required for the component and the method is shown in Fig. 3.18(*b*). There are three basic parts which can be drawn separately and then assembled. The main case profile requires an elliptical shape to be converted into isometric projection. Three ordinates are illustrated as typical examples of the method but more ordinates are required near to the top of the curve for accuracy. Connect the points on the curve with a polyline of 0.4 mm thick and use **Fit** on the **Edit Polyline** feature to improve the profile. The bottom semicircle will appear as an isometric ellipse. The cylinder for the motor cover is best drawn away from the elliptical fan housing and then moved into position after hidden detail has been removed. This reduces the number of construction lines on the same spot.

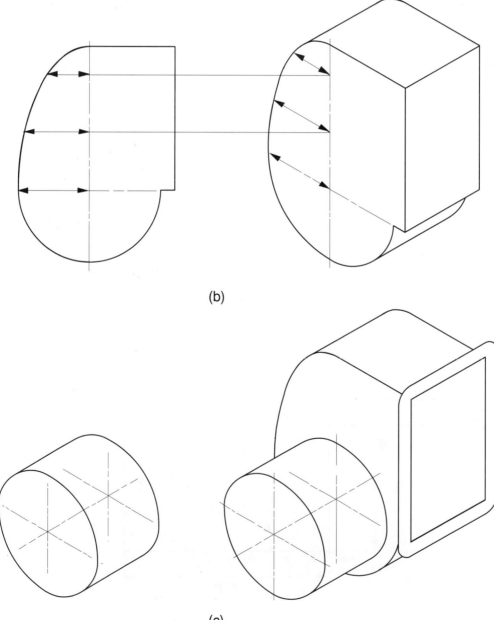

(b)

(c)

Figure 3.18   (continued)

The flange can be drawn *in situ*. This requires a 20 mm diameter isometric circle to be divided into quarters and the parts moved into the four corner positions.

Planometric projection

Planometric projection is a popular form of pictorial drawing using the exact shape of an orthographic plan. Make a drawing of the dovetail joint plan view given in Fig. 3.19(*a*) and rotate it about the corner A though an angle of 60°. Use the COPY command and position a duplicate in the space immediately above. Delete linework to leave the parts

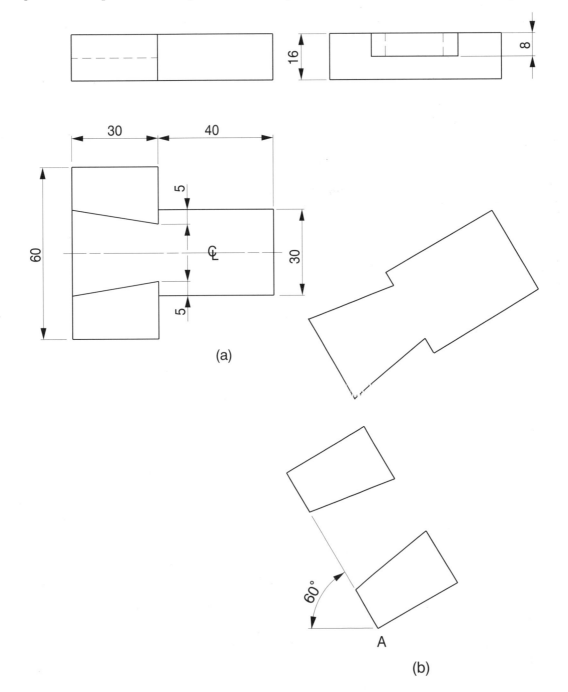

(a)

(b)

Figure 3.19

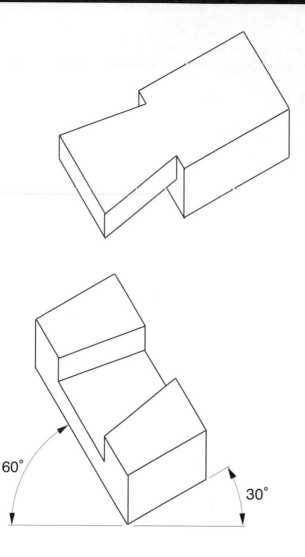

Figure 3.20

shown in Fig. 3.19(*b*). The extra sloping lines in the pictorial views can be added by copying and repositioning each of them 8 and 16 mm below as detailed in the end view. Figure 3.20 shows the completed construction.

Planometric projections are also presented with plan views turned through 45°, and it is a common practice to reduce the vertical heights to a half, two-thirds or three-quarters of the true dimensions in an attempt to improve proportions. In Fig. 3.21 this example has been constructed with the vertical heights changed to three-quarter size for comparison purposes. Planometric presentations are often found in catalogues supplied by kitchen equipment manufacturers.

Part of an extruded moulding is shown in Fig. 3.22(*a*) and Fig. 3.22(*b*) illustrates the construction for a planometric drawing with the given plan rotated through 45° about corner A.

All of the heights from the given front elevation have been halved to improve proportions, and two typical constructional dimensions are shown. Set out the given plan view and use the MOVE command to raise the two rectangular parts by 22.5 mm and 45 mm as shown in Fig. 3.29. Plot each of the curves on the front face and use the COPY feature to

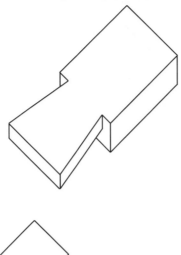

Figure 3.21

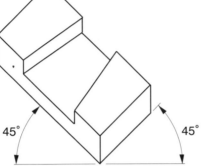

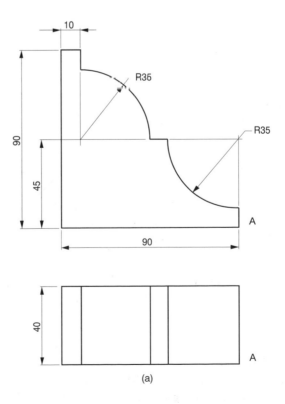

(a)

Figure 3.22

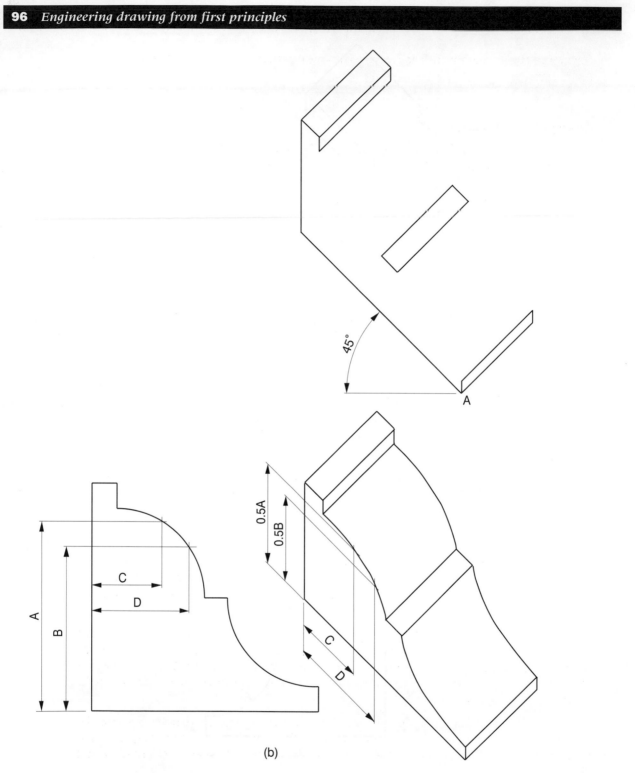

(b)

Figure 3.22 (continued)

reproduce the curves on the rear face. Two typical construction points for the top curve are illustrated.

Simple animation

The COPY and ROTATE features can be used to demonstrate a simple form of animation. Design a bird that you fancy of reasonable size with the CIRCLE and LINE commands. Enlarge the eyeball and beak with the ZOOM command before filling these small areas with a few polyline strokes. Add centre lines where the head, wings and legs are required to rotate.

Using the COPY command, draw boxes in turn, just sufficient to cover each of the component parts of the bird, and these will enable you to reproduce them separately

Figure 3.23

at a convenient position to the right-hand side of the drawing (Fig. 3.23). You will note that the centre lines move with each of the parts, and by superimposing the intersections you will be able to reassemble the bird and also change, if you wish, the profile and the orientation of the limbs. The centre lines act as datums.

Experiment by copying four birds and use the ROTATE command to alter the position of the limbs in turn (Fig. 3.24). Note that the program requires you to draw a box surrounding the limb. Select the rotation point, then manually change the angle. The box may well cover parts of other features, but only complete limbs within the boundaries will be rotated as required on the assembly.

Baby birds look very much like their mothers and can easily be reared by using the MIRROR feature, then the SCALE feature to make them younger. EDIT the shapes to give finished profiles. A lot of fun can be had by changing feature shapes, orientation and sizes.

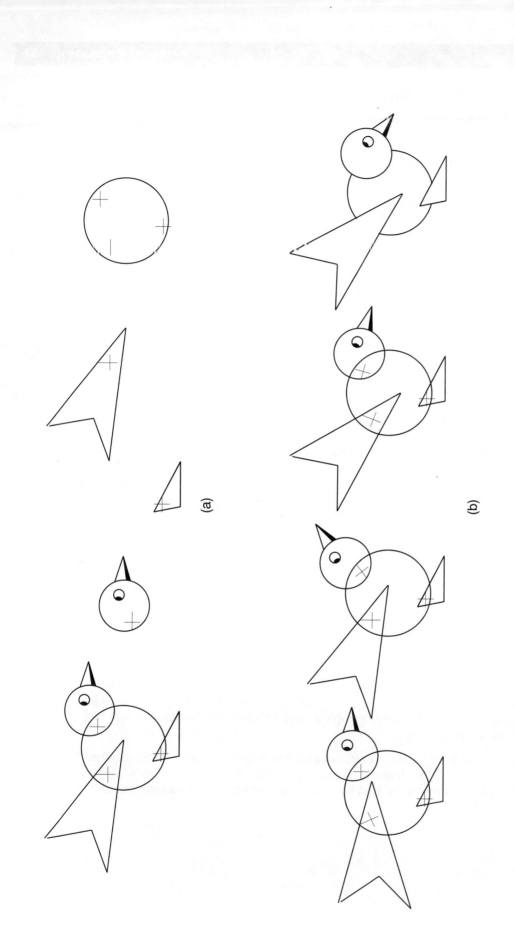

(a)

(b)

Figure 3.24

Chapter 4

Orthographic projection

This chapter introduces:

- First angle or European projection.
- Third angle or American projection.
- Projection examples and the selection of views.
- Standard projection symbols.

A pictorial view of a cable connector is shown in Fig. 4.1 and detailed dimensioned drawings are required for its manufacture. Draughtsmen use a multi-view orthographic projection system to present this information. Two methods are in common use and both are acknowledged by British and International Standards as acceptable and of equal merit. First angle or European projection is probably more widely used in Europe and is normally used in Standards here for uniformity of presentation. Third angle is the American projection method and in regular use by industry here. The professional draughtsman needs to be familiar with both systems. Drawings are not only produced for manufacturing purposes but also to communicate information for readers in other

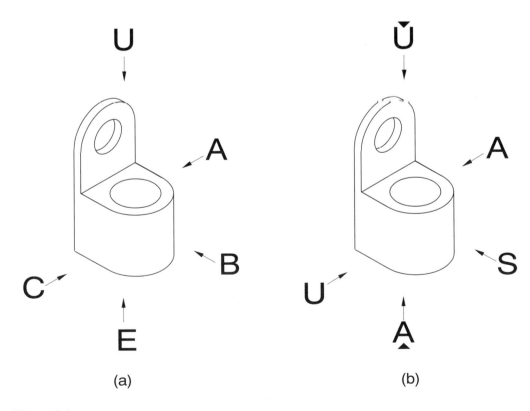

(a)   (b)

Figure 4.1

professions. Drawings may be used as legal documents. The drawing must therefore be clear, accurate, concise and free of any ambiguity. To assist interpretation of multi-view drawings, the Standards state that the drawing sheet should contain a symbol or a note to state which projection has been used.

The symbols are illustrated in Fig. 4.2 and their significance will be appreciated after the following exercise.

Cut a cube with sides of about 30 mm sides from polystyrene, which can generally be found as a packing material, and draw the letters A, B, C, E and U on five of the faces to agree with the orientation shown pictorially on Fig. 4.3. Place the cube on a flat surface with the letter B facing towards you. Roll the cube over to your left and the letter A will appear. Start again and roll the block to your right and C appears. Again from the centre, roll the block away from you to show the E and finally, again from the centre, roll the block towards you and the U is visible. The five faces of the cube are drawn in first angle orthographic projection. Study the pictorial views of the connector in Fig. 4.1(a). You will appreciate that there are a variety of ways to present the illustration. Before commencing

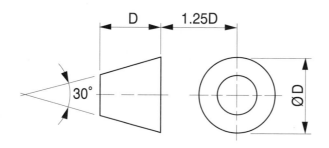

SYMBOL PROPORTIONS

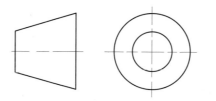

FIRST ANGLE
PROJECTION

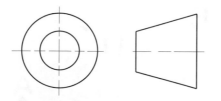

THIRD ANGLE
PROJECTION

Figure 4.2

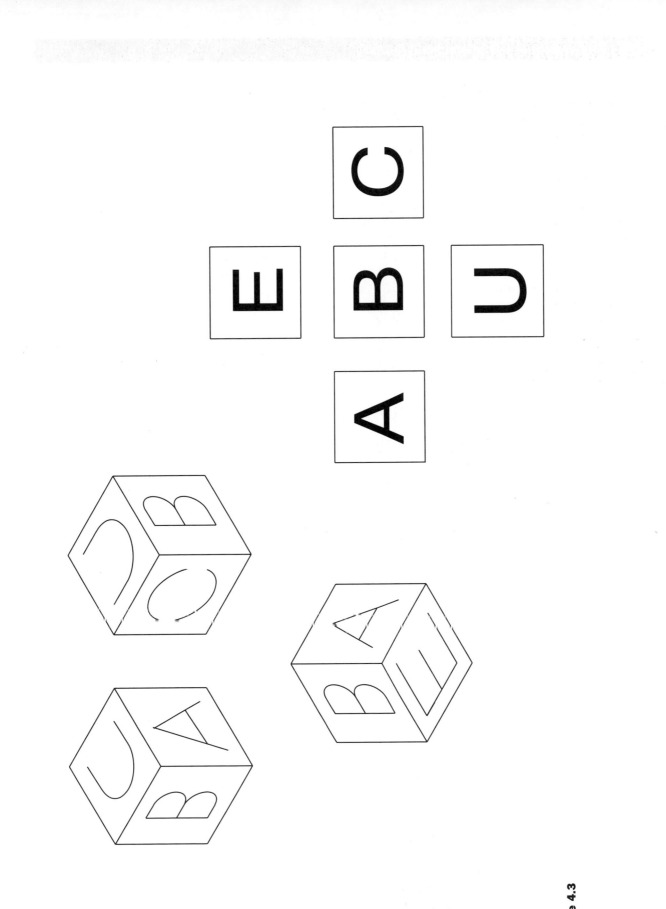

Figure 4.3

a drawing, the draughtsman must make a decision to select a suitable front view as their job is to present the component in the clearest possible manner to convey its shape and form to the reader. There are five possible viewing directions that are square to the faces of the component. An addition view can be taken from the rear if required. You can look at any other angle from space and draw auxiliary projections, but the geometrical constructions become involved. Some examples are provided later in the text.

Having made an arbitrary decision to start with view B at the front then the other projections follow and are illustrated in Fig. 4.4. The draughtsman only draws sufficient views to clarify the job in hand. A study of Fig. 4.4 clearly shows that the left-hand and right-hand views are very similar and only one need be kept. The views above and below are also very similar, but the view below is chosen as it shows the inside corner in a full line instead of the hidden line in the view above. The finished dimensioned detail drawing is shown in Fig. 4.5.

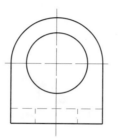

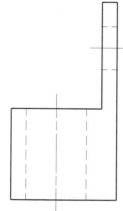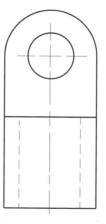

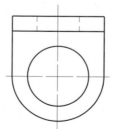

Figure 4.4

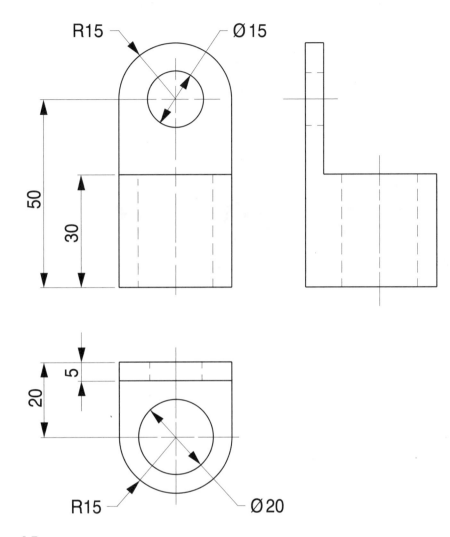

Figure 4.5

American or third angle projection is demonstrated with the other cube marked as indicated on Fig. 4.6. Start, as before, with the letter S in the centre and note that you need to lift the cube, rotate it through 90° then place it on the left side of the S to obtain letter U. Repeat from the centre by lifting the cube, rotate in the opposite direction and place it on the right side for letter A. Check the procedure with the top and bottom views. The five views in third angle projection are drawn in Fig. 4.7. A comparison of both of these presentations will show the differences between them.

In first angle the projected view drawn on the right-hand or opposite side of the front view shows the left-hand face. In third angle projection the view on the left side shows the adjacent left-hand face.

In first angle the plan view which is drawn beneath the front view shows the top face of the component, i.e. a bird's eye view. In third angle the plan view of the top of the component is drawn above the component.

Check how the principles are applied to the other sides. Study Fig. 4.4 and Fig. 4.7: we started with identical front views in both projections. The position of the other two pairs of end views and plans are exchanged.

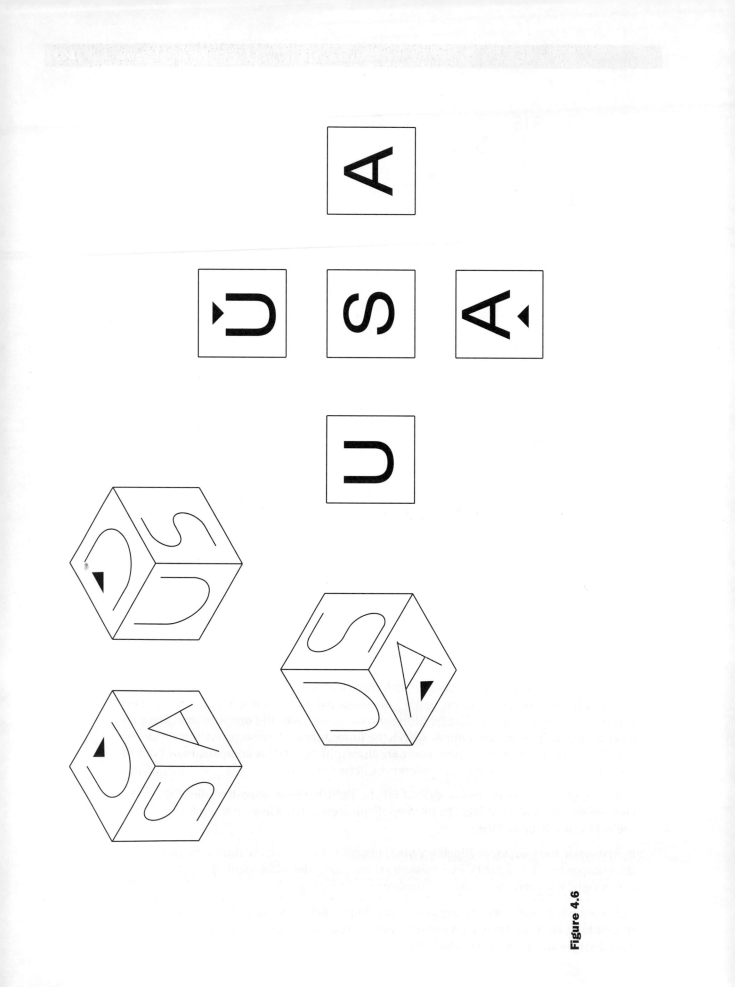

Figure 4.6

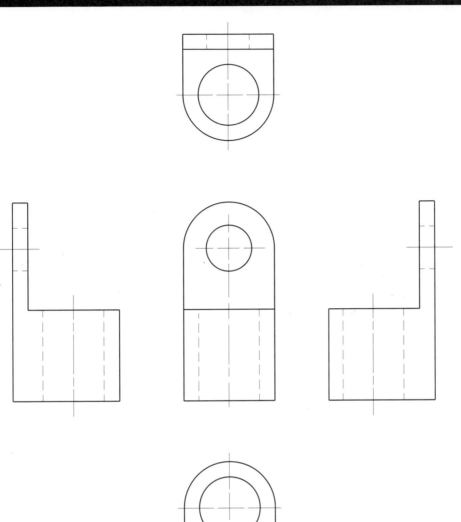

Figure 4.7

Remember you can draw in first or third angle projection but must state which one you have used. What you cannot do is draw three views of a component where the projections are mixed with one in first and the other in third angle, since some features can be wrongly positioned and will not conform with design intentions. Please see a later note.

You will find that in industry the words 'views' and 'elevations' are widely used. We freely refer to 'plans' with reference to drawings of all parts of buildings and therefore must accept broad interpretations of these terms.

First angle projection exercises

Four isometric drawing examples were given previously in Fig. 3.3 and drawn on a grid of equilateral triangles with 10 mm sides. Figure 4.8(a)–(d) show five first angle orthographic views for each of the examples, assuming that the arrows 1, 2, 3, and 4 point to the front views. These exercises are designed to rearrange selected parts from existing drawings to make new drawings.

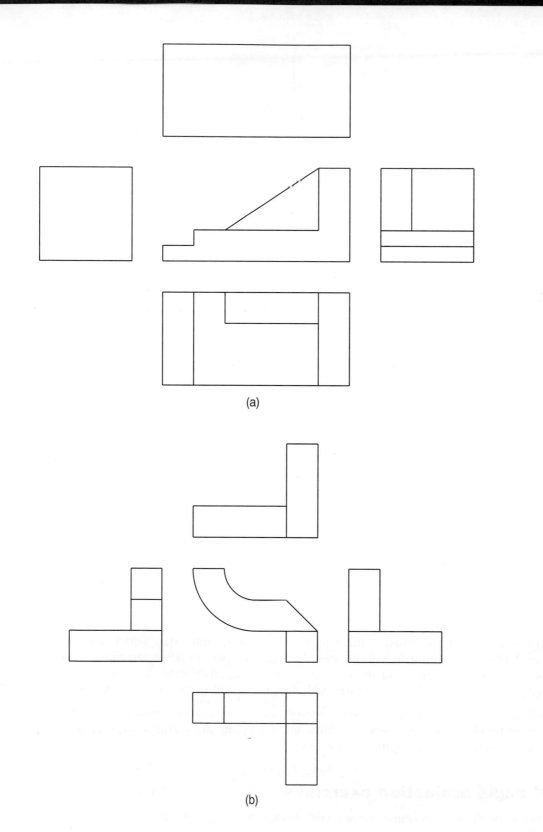

(a)

(b)

Figure 4.8

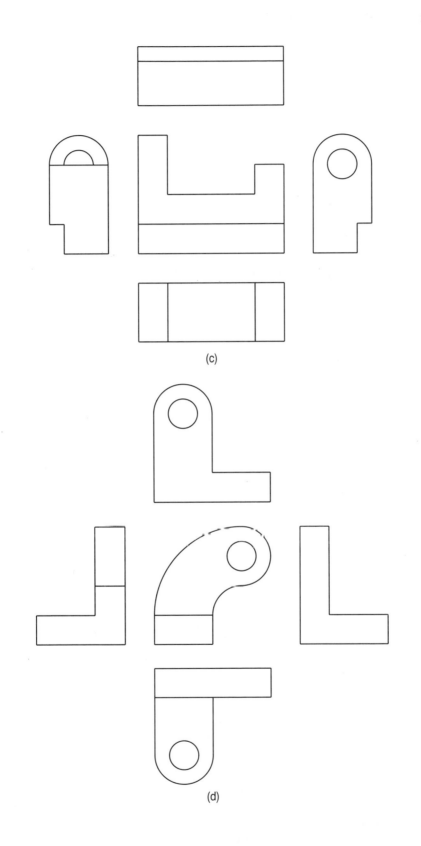

(c)

(d)

Figure 4.8 (continued)

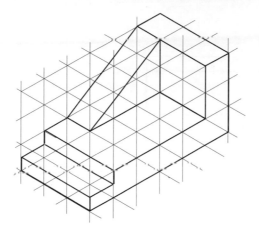

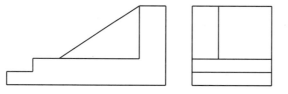

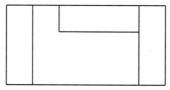

FIRST ANGLE SOLUTION

Figure 4.9

A DXF file (drawing interchange format) allows you to save a current drawing in a format that can be read later and inserted into another file. Figure 4.9 consists of parts of three previously drawn exercises, namely

1. the top left component on Fig. 3.3;
2. parts of the orthographic views at the bottom of Fig. 4.8 (*a*);
3. the first angle symbol from Fig. 4.2.

In order to combine three drawings we need one basic drawing and then transfer onto it the other two parts. For this example we will transfer sections 1 and 3 above onto a part of Fig. 4.8(*a*).

Open your file for Fig. 3.3 and rename a copy with your own new file name (say ZZZ). Delete the three parts you do not need. In the **File** menu click on **Import/Export**. A sub-menu will appear. Choose **DXF Out** and the Create DXF File dialogue box appears on the screen. Insert your drawing file name and click **OK**. You will now have a DXF file of the isometric view named ZZZ.DXF.

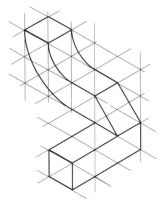

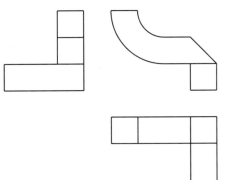

FIRST ANGLE SOLUTION

Figure 4.10

Open your file for Fig. 4.2 and save a copy with a new file name (say XXX). Erase everything except the required first angle symbol. Repeat the above procedure to make a DXF file of the symbol, which you will be able to use for this exercise and other applications. Open your file containing Fig. 4.8(a) and save it under a separate file number (say YYY). Erase the two blank views and leave the three orthographic views to the right-hand part of the screen. The reason is simply that when the DXF files are inserted, you need them to appear on an unoccupied part of the screen.

Click on **Import/Export** again in the **File** menu followed by **DXF In**. Insert your DXF file number for the isometric drawing in the dialogue box and your previous drawing will now appear. Repeat the procedure for the symbol and insert this drawing.

You now have on the screen the component parts of Fig. 4.9. Rearrange them using the Move command and add the first angle solution text. This procedure is regularly used so repeat the exercise for Fig. 4.10 to become proficient in the operation.

Chapter 5

Text and dimensions

This chapter introduces you to:

- Text types and editing options.
- Special characters.
- Overscoring and underscoring.
- Dimensioning principles.
- Functional and Auxiliary dimensions.
- Toleranced dimensions.
- The selection of limits and fits from ISO tables with shaft and hole combinations.
- Combining characters to produce limits and fits for use on drawings.

Text

There are many different fonts available in CAD but BS 308 favours types which are open form and free from serifs and embellishments. Capital letters are preferred. Vertical or sloping letters are acceptable but for ease of application the vertical form is the most convenient for CAD. It is not acceptable to mix styles. Examples are shown in Fig. 5.1.

To be consistent, all the text on the drawings in this book has been added using the Roman Duplex style. Select **Roman Duplex** from the **Text Style** dialogue box in the **Settings** menu and click **OK**. Various text options are now available at the command line as follows.

```
Height <0.000>: Enter 3.5.
Width factor <1.000>: Enter 0.9.
Obliquing angle <0>: OK.
Backwards?<N>: OK.
Upside down?<N>: OK.
Vertical?<N>: OK.
```

(a) ABCDEFGHIJKLMNOPQRSTUVWXYZ
abcdefghijklmnopqrstuvwxyz
1234567890

(b) *ABCDEFGHIJKLMNOPQRSTUVWXYZ*
abcdefghijklmnopqrstuvwxyz
1234567890

(c) ABCDEFGHIJKLMNOPQRSTUVWXYZ
aabcdefghijklmnopqrstuvwxyz
1234567890

(d) *ABCDEFGHIJKLMNOPQRSTUVWXYZ*
aabcdefghijklmnopqrstuvwxyz
1234567890

Figure 5.1

The command line reads Romand is now the current text style.

Choose **Text** from the **Draw** menu and position the cursor box where the text is required on the screen. The characters that you type will also appear at the command line and should you subsequently use the MOVE command all the inserted characters will move together as a block. Each time the Text prompt reappears, DTEXT draws a cursor box and the next message can be inserted.

If, for instance, you have a message covering two lines which you wish to centralise, then the two lines must be inserted separately so that you can move one independently of the other. This is often the case where dimensioning notes are involved, and many examples appear in the book.

Special characters

You will find it necessary to use the symbol ϕ preceding the numeral for a diameter. The symbol here from the word processor is lowercase and some CAD programs do not give an uppercase version. With my software the symbol can be obtained by inserting **%%c** followed by the appropriate number and pressing the <Enter> key twice, but the result is also lowercase. I have seen both upper and lowercase symbols on drawings, but personally prefer the uppercase type. This need not be a problem. If I need to insert a dimension of diameter 40 on the drawing I use the TEXT command and write **O40**. Then a separate / sign can be repositioned using the MOVE command. This provides a neat, wholly uppercase dimension.

For an angular degree sign on the drawing, e.g. 50°, it is necessary to type **50%%d** and press the <Enter> key twice, or **35%%d C** for 35° Centigrade. The plus or minus tolerance sign is obtained from inserting **%%p** and the percent sign requires **%%%**, plus <Enter> twice.

Overscoring and underscoring can be arranged and requires **%%o** or **%%u** followed by the numeral. The actual position of the horizontal line is not however, positioned quite centrally by the software. A centrally positioned fraction may be drawn in three steps by drawing the numbers separately plus an oblique line. The line is then rotated to the horizontal position and the group rearranged with the MOVE command.

You will find these separate manipulations of considerable value during dimensioning applications. Figure 5.2 shows step by step examples for practice.

Dimensioning

BS 308 gives a complete coverage and explanation of the principles of dimensioning and tolerancing of size. Standards relating to this subject are always subject to review.

Please add dimensions to all the drawing exercises as you work your way through this book. They are models in themselves of acceptable methods of applied dimensioning, and hopefully you will learn by example. You will also increase your draughting speed and ability to comprehend basic engineering details. Please also, therefore, accept the following additional notes to those demonstrated in the progressive exercises through the book.

Dimensions are applied to a drawing with associated tolerances, where necessary, with information to define the object completely. A dimension for a particular feature should appear only once and be placed on the drawing where that feature is most clearly shown.

Features are the elements of objects and shapes, and the term applies to lengths,

056	60.00	30.12
/	60.04	30.00
Ø56	0.4	0
	±	/
ø56	60±0.4	Ø 30.12
	60	30.00

50°	35°C	100%

$\dfrac{15}{32}$	$\dfrac{15}{\overline{32}}$	$\dfrac{5}{16}$	$\dfrac{5}{\overline{16}}$	$\dfrac{3}{4}$	$\dfrac{3}{4}$
—		—		—	

Figure 5.2

diameters, widths, etc. A particular feature which is vital to the performance of a component is called a 'functional feature'. An example could be the diameter of a spindle or the diameter of a locating peg. The associated dimension is known as a 'functional dimension'.

Auxiliary dimensions are given for information only. They are additional to dimensions used in manufacturing and are shown in brackets.

A designer will use preferred sizes where possible. These can be considered to relate to the sizes of standard tools, e.g. drills, taps, dies and raw material stocks.

Dimensions are normally stated in millimetres, but the unit symbol 'mm' may be omitted. The drawing sheet should have a note stating the units used. Dimensions are quoted on drawings to the least number of significant figures, e.g. 25 and not 25.0. The decimal marker should be bold, given a full letter space and placed on the line. Dimensions less than 1 should be preceded by 0, e.g. 0.75. Angular dimension may be quoted on drawings either in degrees, minutes and seconds or in degrees and decimals of a degree, e.g. 25°; 25°30' or 25.5°; 25°30'30" or 25.55°.

The main components of a dimension are the projection lines, the dimension lines, the arrows and the appropriate numbers and notes. In order to create a style of your own it is necessary to keep to certain standard practices and repeat them. Most drawings form part of a set, for example assemblies and their associated component details, so the general presentation must be consistent.

All CAD programs offer various options relating to dimensioning practice and the application of the dimensions to the drawing. At this stage, perhaps it would be helpful to take a drawing and point out certain aspects regarding layout and positioning.

Fig. 5.3 shows a small plate with various features. The plate would be manufactured by cutting and machining a rectangular piece of material of size 65 × 80 mm. The various features on the plate are positioned from the sides and each side would be expected to be straight and square to the surface. All surfaces should have a good finish.

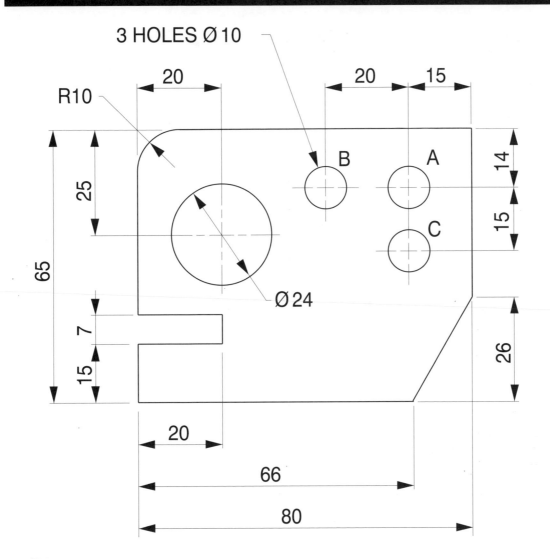

Figure 5.3

Consider hole A near the top right-hand corner of the plate in Fig. 5.3 and its position from the top and right-hand sides. For these measurements the two sides are considered to be datums. The centre lines of hole A are used to position the centres of holes B and C, so the centre lines of hole A are also regarded as datums for this purpose. Assume that the inspector will pass work where each dimension is within a tolerance of ±1mm.

In case 1 (Fig. 5.4) all four dimensions are at their lower limits and in case 2 the dimensions are given at their higher limits. Case 3 gives the condition where hole A is at the low limit of size and holes B and C are at their high limit with respect to hole A.

All three conditions will pass inspection if the dimensions are applied in this way. It follows that the position of hole B from the right-hand side measured horizontally may be between 33 and 37mm. The centre of hole C measured vertically from the top edge may be between 27 and 31mm.

Case 4 shows a different result where hole B will be positioned between 34 and 36mm from the right side and hole C between 28 and 30mm. This method of dimensioning

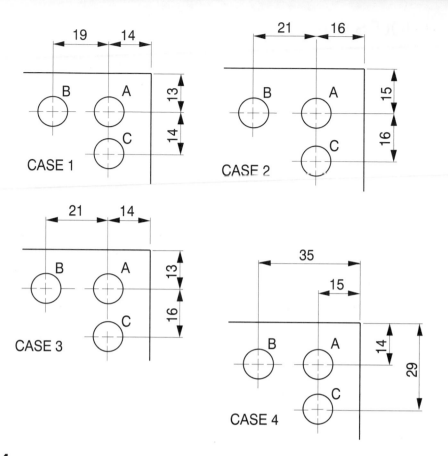

Figure 5.4

eliminates the cumulative errors arising from the previous 'chain' dimensioning in cases 1, 2 and 3.

Notice that the dimensions for the angle cut at the bottom right-hand corner imply that the measurements are taken from the left side and the bottom of the plate. Only the minimum dimensions required to define a shape are necessary, and in this case to quote an angle in addition would be superfluous. If an angle was given then one of the distances would not be necessary.

It is recommended that dimensions are read from the bottom of the drawing or by turning the drawing clockwise so that the right side is at the bottom. Dimensions should be positioned centrally above the dimension line. Larger dimensions are placed outside smaller dimensions. It is considered good practice to place dimensions outside the drawing as first choice if space is restricted, but this is not always possible, especially with small radii and detail in the centre of a large drawing area. Copy Fig. 5.3 exactly as an example of dimensioning practice and add the given details with the dimension text 3.5 mm in height. If dimensions are positioned 1 mm above the dimension line at the centre point, then the side of the 4 × 2 mm arrow can be used to line up the bottom of the number.

When copying drawings of this type, set out the shape using a GRID of 10 mm and a SNAP setting of 1 mm. The component outline and details are completed with a 0.4 mm poly-line. In the case of holes A, B and C, draw one hole and its centre line and copy the others.

Position dimension lines 10 mm apart. The projection lines from the component should be inserted leaving a 1 mm gap at the drawing outline and extending past the arrow tip by 1 mm. Any small overruns with lines can easily be tidied up using the BREAK feature. Likewise, lines can easily be added accurately where necessary if the SNAP feature is used.

Draw a single arrow 4 mm long and 2 mm wide using the SOLID option in the Draw menu making sure that SNAP is engaged. Copy the arrow three times and use ROTATE three times to obtain the four arrows as shown. I simply copy them wherever they are needed and snap them into position. The arrows for sloping dimensions are copied into the vicinity then oriented with the ROTATE option and positioned exactly where required with the MOVE feature. I personally find that to be in complete separate command of the arrow, text, dimension and projection lines is very desirable for editing and repositioning dimensioning details on original work.

I also find it quicker to place in a space away from the drawing a text list of numbers and notes I require and then rotate as necessary and move them into position accurately, taking advantage of the ZOOM feature.

Fig. 5.5 shows a circular disc with six holes equally spaced around a pitch circle diameter of 70 mm. Copy the disc by setting out one hole with its symmetrically spaced centre lines. The circular arc can easily be broken at selected points on either side of the vertical. Draw one hole profile and edit the line to 0.4 mm thickness. Use the POLAR ARRAY feature to add the other five holes.

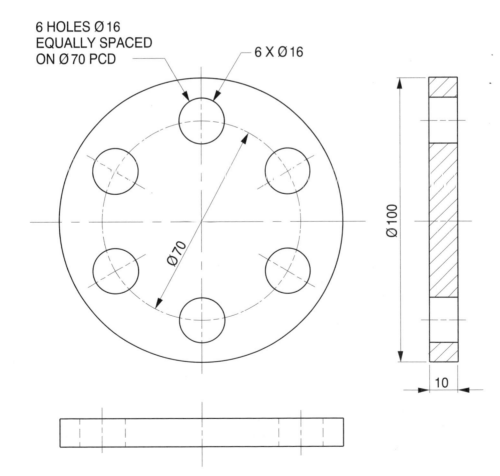

Figure 5.5

Table 5.1 Grades of fit for combinations of shafts and holes

Extracted from BS 4500:1969

BRITISH STANDARD

Data Sheet 4500A
Issue 1 February 1970
Confirmed August 1985

SELECTED ISO FITS—HOLE BASIS

Diagram to scale for 25 mm diameter

Clearance fits: H11/c11, H9/d10, H9/e9, H8/f7, H7/g6, H7/h6
Transition fits: H7/k6, H7/n6
Interference fits: H7/p6, H7/s6

Legend: Holes / Shafts

All tolerance values in units of 0.001 mm (upper / lower).

Nominal sizes Over (mm)	To (mm)	H11	c11	H9	d10	H9	e9	H8	f7	H7	g6	H7	h6	H7	k6	H7	n6	H7	p6	H7	s6
–	3	+60/0	−60/−120	+25/0	−20/−60	+25/0	−14/−39	+14/0	−6/−16	+10/0	−2/−8	+10/0	0/−6	+10/0	+6/0	+10/0	+10/+4	+10/0	+12/+6	+10/0	+20/+14
3	6	+75/0	−70/−145	+30/0	−30/−78	+30/0	−20/−50	+18/0	−10/−22	+12/0	−4/−12	+12/0	0/−8	+12/0	+9/+1	+12/0	+16/+8	+12/0	+20/+12	+12/0	+27/+19
6	10	+90/0	−80/−170	+36/0	−40/−98	+36/0	−25/−61	+22/0	−13/−28	+15/0	−5/−14	+15/0	0/−9	+15/0	+10/+1	+15/0	+19/+10	+15/0	+24/+15	+15/0	+32/+23
10	18	+110/0	−95/−205	+43/0	−50/−120	+43/0	−32/−75	+27/0	−16/−34	+18/0	−6/−17	+18/0	0/−11	+18/0	+12/+1	+18/0	+23/+12	+18/0	+29/+18	+18/0	+39/+28
18	30	+130/0	−110/−240	+52/0	−65/−149	+52/0	−40/−92	+33/0	−20/−41	+21/0	−7/−20	+21/0	0/−13	+21/0	+15/+2	+21/0	+28/+15	+21/0	+35/+22	+21/0	+48/+35
30	40	+160/0	−120/−280	+62/0	−80/−180	+62/0	−50/−112	+39/0	−25/−50	+25/0	−9/−25	+25/0	0/−16	+25/0	+18/+2	+25/0	+33/+17	+25/0	+42/+26	+25/0	+59/+43
40	50	+160/0	−130/−290	+62/0	−80/−180	+62/0	−50/−112	+39/0	−25/−50	+25/0	−9/−25	+25/0	0/−16	+25/0	+18/+2	+25/0	+33/+17	+25/0	+42/+26	+25/0	+59/+43
50	65	+190/0	−140/−330	+74/0	−100/−220	+74/0	−60/−134	+46/0	−30/−60	+30/0	−10/−29	+30/0	0/−19	+30/0	+21/+2	+30/0	+39/+20	+30/0	+51/+32	+30/0	+72/+53
65	80	+190/0	−150/−340	+74/0	−100/−220	+74/0	−60/−134	+46/0	−30/−60	+30/0	−10/−29	+30/0	0/−19	+30/0	+21/+2	+30/0	+39/+20	+30/0	+51/+32	+30/0	+78/+59
80	100	+220/0	−170/−390	+87/0	−120/−260	+87/0	−72/−159	+54/0	−36/−71	+35/0	−12/−34	+35/0	0/−22	+35/0	+25/+3	+35/0	+45/+23	+35/0	+59/+37	+35/0	+93/+71
100	120	+220/0	−180/−400	+87/0	−120/−260	+87/0	−72/−159	+54/0	−36/−71	+35/0	−12/−34	+35/0	0/−22	+35/0	+25/+3	+35/0	+45/+23	+35/0	+59/+37	+35/0	+101/+79
120	140	+250/0	−200/−450	+100/0	−145/−305	+100/0	−85/−185	+63/0	−43/−83	+40/0	−14/−39	+40/0	0/−25	+40/0	+28/+3	+40/0	+52/+27	+40/0	+68/+43	+40/0	+117/+92
140	160	+250/0	−210/−460	+100/0	−145/−305	+100/0	−85/−185	+63/0	−43/−83	+40/0	−14/−39	+40/0	0/−25	+40/0	+28/+3	+40/0	+52/+27	+40/0	+68/+43	+40/0	+125/+100
160	180	+250/0	−230/−480	+100/0	−145/−305	+100/0	−85/−185	+63/0	−43/−83	+40/0	−14/−39	+40/0	0/−25	+40/0	+28/+3	+40/0	+52/+27	+40/0	+68/+43	+40/0	+133/+108
180	200	+290/0	−240/−530	+115/0	−170/−355	+115/0	−100/−215	+72/0	−50/−96	+46/0	−15/−44	+46/0	0/−29	+46/0	+33/+4	+46/0	+60/+31	+46/0	+79/+50	+46/0	+151/+122
200	225	+290/0	−260/−550	+115/0	−170/−355	+115/0	−100/−215	+72/0	−50/−96	+46/0	−15/−44	+46/0	0/−29	+46/0	+33/+4	+46/0	+60/+31	+46/0	+79/+50	+46/0	+159/+130
225	250	+290/0	−280/−570	+115/0	−170/−355	+115/0	−100/−215	+72/0	−50/−96	+46/0	−15/−44	+46/0	0/−29	+46/0	+33/+4	+46/0	+60/+31	+46/0	+79/+50	+46/0	+169/+140
250	280	+320/0	−300/−620	+130/0	−190/−400	+130/0	−110/−240	+81/0	−56/−108	+52/0	−17/−49	+52/0	0/−32	+52/0	+36/+4	+52/0	+66/+34	+52/0	+88/+56	+52/0	+190/+158
280	315	+320/0	−330/−650	+130/0	−190/−400	+130/0	−110/−240	+81/0	−56/−108	+52/0	−17/−49	+52/0	0/−32	+52/0	+36/+4	+52/0	+66/+34	+52/0	+88/+56	+52/0	+202/+170
315	355	+360/0	−360/−720	+140/0	−210/−440	+140/0	−125/−265	+89/0	−62/−119	+57/0	−18/−54	+57/0	0/−36	+57/0	+40/+4	+57/0	+73/+37	+57/0	+98/+62	+57/0	+226/+190
355	400	+360/0	−400/−760	+140/0	−210/−440	+140/0	−125/−265	+89/0	−62/−119	+57/0	−18/−54	+57/0	0/−36	+57/0	+40/+4	+57/0	+73/+37	+57/0	+98/+62	+57/0	+244/+208
400	450	+400/0	−440/−840	+155/0	−230/−480	+155/0	−135/−290	+97/0	−68/−131	+63/0	−20/−60	+63/0	0/−40	+63/0	+45/+5	+63/0	+80/+40	+63/0	+108/+68	+63/0	+272/+232
450	500	+400/0	−480/−880	+155/0	−230/−480	+155/0	−135/−290	+97/0	−68/−131	+63/0	−20/−60	+63/0	0/−40	+63/0	+45/+5	+63/0	+80/+40	+63/0	+108/+68	+63/0	+292/+252

BRITISH STANDARDS INSTITUTION, 2 Park Street, London, W1A 2BS

SBN: 580 05766 6

Notice in the plan view that if hidden detail is required, then the dotted lines for the holes are oriented about the projection of the pitch circle. The end view here shows a sectional view, assuming the disc to be cut along the vertical centre line. The three areas where material is cut are shown cross-hatched at 45°. Two methods of dimensioning the holes are given.

Full instructions may be stated as shown. If spacings are self-evident then the note 'equally spaced' may be omitted. Where a number of holes of the same size are drawn on the same view, then the short note here is acceptable. In this case the size of the pitch circle is given separately.

Limits and fits

Many components, such as some garden tools, work in a completely satisfactory manner after manufacture to an acceptable standard where tight dimensional controls are not necessary. However, production to a higher degree of precision is needed for more sophisticated assemblies where dimensional accuracy is vital for successful operation in service.

It is taken for granted that if a component in, for example, a washing machine or a car is broken or worn out, you will be able to replace it with a spare part. Mass production techniques are possible because components in an assembly are interchangeable and the designer selects suitable dimensions to ensure that an assembly functions in the intended manner.

Since exact measurements cannot be made, the designer selects, when necessary, acceptable high and low limits of size for each important feature dimension. If the component is manufactured to a size somewhere between them, then it will be passed by the inspector. Dimensions for high and low limits are each checked by 'go' and 'no go' inspection gauges to ensure that the final product size is not too big or not too small.

Table 5.2 Shaft and hole combinations

Mating assembly		BS 4500 class	Interpretation and type of work
Hole	Shaft		
H11	c11	Clearance	Very loose running
H9	d10		Loose running
H9	e9		Easy running
H8	f7		Running fit; good-quality fit between bearing and rotating spindle
H7	g6		Close running
H7	h6		Sliding fit for non-running assemblies such as splined shafts
H7 fits	k6	Transition	Easy push; recommended for location with virtually no clearance Dowels and inner rings of ball bearings
H7	n6		Push fit for tight assemblies
H7	p6	Interference	Light press with slight interference Allows parts to be dismantled without overstraining
H7	s6		Press fit with definite interference Bearing bush assembly

Measurements are taken from clearly defined datum axes, planes or surfaces identified on the drawing. The differences between high and low limits vary according to the size of work and the degree of precision required. The mathematical difference between high and low limits is called the 'tolerance'. Toleranced dimensions for very precise and accurate work invariably increase production costs. Shape and form are controlled by the application of limits and fits. British Standards have made recommendations for tolerances to provide different grades of fits for various combinations of shafts and holes and extracts are given in Table 5.1 shown later with typical examples.

Mass production methods require manufacturers to supply interchangeable parts, but remember that if many identical components are required, we only need one drawing for each component. Thus dimensions must be applied with care in a clear unambiguous manner.

Table 5.2 shows the ISO recommendations for different combinations. The complete ISO system of limits and fits gives a large number of combinations, but in practice the majority of fits can be provided by the selection given in Table 5.2.

Note the term 'hole basis' at the head of the table. This system of fits is designed to associate different shaft sizes with a single hole. You will appreciate that with a hole obtained from a 25 mm drill, for instance, it is difficult to vary its size after it has been manufactured. If a peg of 25 mm nominal diameter is to be fitted into the hole, then for an easier fit it is more convenient to grind a little off the male part than to increase the size of the hole.

Applications in heavy engineering often do require various diameters to fit shafts of standard size and there is also a 'shaft basis' system available.

Clearance fits always provide a clearance which means that the shaft is always smaller than the hole. They are often referred to as running fits. Interference fits occur where the shaft is always of a larger diameter than the hole and pressure is required to force

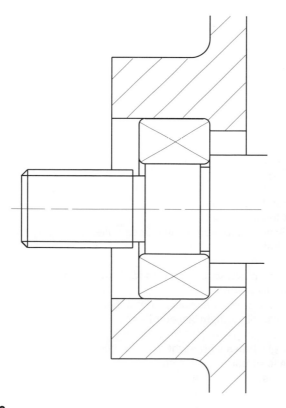

Figure 5.6

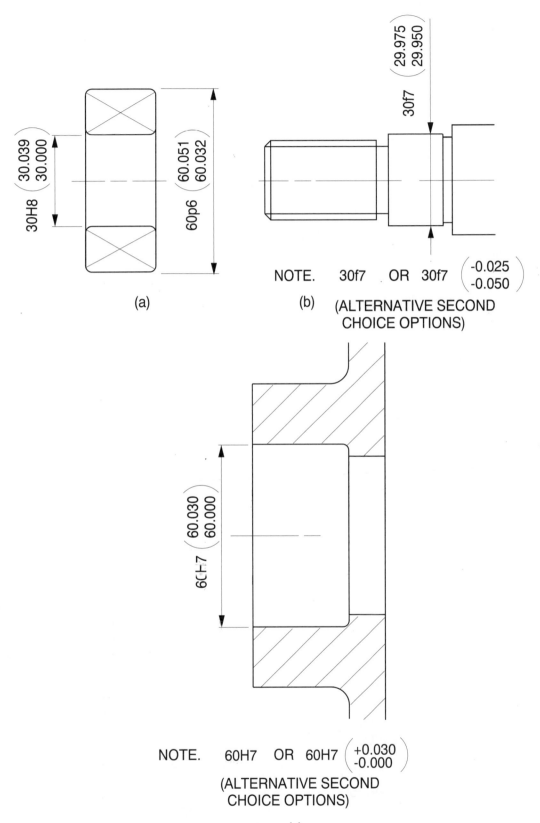

(a)

NOTE.   30f7   OR   30f7 $\begin{pmatrix} -0.025 \\ -0.050 \end{pmatrix}$

(b)   (ALTERNATIVE SECOND
CHOICE OPTIONS)

NOTE.   60H7   OR   60H7 $\begin{pmatrix} +0.030 \\ -0.000 \end{pmatrix}$

(ALTERNATIVE SECOND
CHOICE OPTIONS)

(c)

Figure 5.7

assembly. It is often referred to as a drive fit. Alternatively, before assembly, the component with the hole may be heated until expansion takes place and then the shaft is inserted; this is often known as a shrink fit. Transition fits occur where there is a small overlap in the tolerance zones and only light pressure is necessary for assembly, this is often known as a push fit.

The designer selects the appropriate combinations and is always aware that as increased precision is required, so costs increase to obtain it. The product designer may apply the selected tolerances with the nominal dimension and the grade of fit on the design drawing. A production draughtsman responsible for issuing a production drawing to the workshop will only need to show the high and low limits of size and should not expect the craftsman to perform mental arithmetic. There are therefore several ways of quoting toleranced dimensions on a drawing and the next exercise is given for this purpose. Assume suitable sizes where other dimensions are not given.

Figure 5.6 shows part of a bearing in its housing at the end of a shaft. The bearing could be a ball or roller bearing, and in a sectional view there would no need to draw the bearing

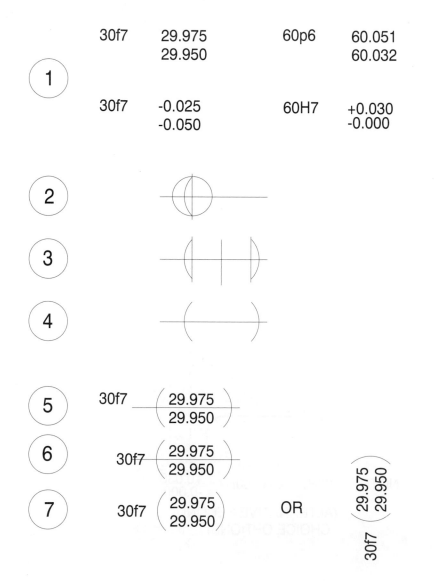

Figure 5.8

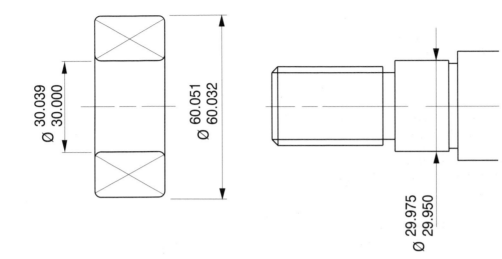

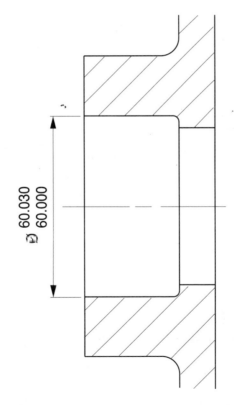

Ø 30.039 / 30.000

Ø 60.051 / 60.032

Ø 29.975 / 29.950

Ø 60.030 / 60.000

Figure 5.9

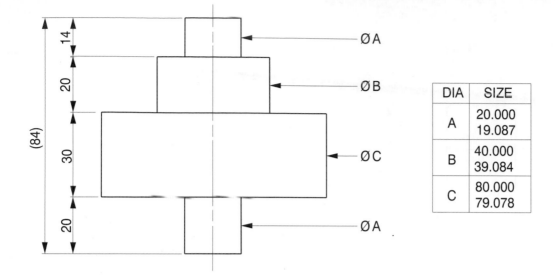

DIA	SIZE
A	20.000
	19.087
B	40.000
	39.084
C	80.000
	79.078

Figure 5.10

internal detail as the BS convention for both types is crossed diagonal lines. Assuming that the manufacturers supplied the bearing and recommended a fit of H7/p6 for the housing recess and H8/f7 on the shaft, then the example shows the first-choice method of dimensioning for the feature sizes. Alternatively, the dimensions 60H7 and 30f7 could have been substituted or the dimensions shown with the relevant tolerances, instead of the feature sizes in the brackets (Fig. 5.7).

Please note that bearing manufacturers issue general catalogues listing dimensional details of their products. Bearings are supplied as completely finished components manufactured to international agreed standards. It is a useful CAD manipulation exercise to set out toleranced dimensions neatly.

Copy the numerical figures shown in Fig. 5.8, using the following procedure.

1. Type separately the lines of text on the screen which need to be put in brackets.
2. You need to draw the large brackets which are symmetrical, so commence with one arc using the three-point option.
3. Draw a mirror line a short distance away and obtain the mirror copy; the distance apart is not important.
4. Delete unwanted lines.
5. Move one side of the bracket arcs to the required distance from the dimensions.
6. Move the single dimension onto the centre line between the other pair.
7. Move the other arc horizontally into position so that spacings are equal and delete the centre construction line.
8. Move into position on detail drawing as required (Fig. 5.9).

Many automatic dimensioning systems are perfectly satisfactory until you try to edit small parts of them, so it is very useful to be able to design and manipulate your own.

Figure 5.10 shows a component with concentric and repeated diameters. A table of finished sizes is an acceptable method of dimensioning. Copy the drawing and set out the table, ensuring that all the separate parts are symmetrically spaced. The overall length of the component is given in brackets for information only and is an example of an auxiliary dimension.

Chapter 6

Three-dimensional projection exercises

This chapter introduces examples where three-dimensional components need to be presented on the screen with sufficient views to define their shape accurately. At the end of the chapter you will be able to project relevant linework between views in order to complete them. Examples show component views in line vertically and horizontally and introduce the projection of curves, tapers and angular surfaces. Conic sections are given to demonstrate constructions for the hyperbola, parabola and ellipse. Intersections are shown between parts of solids joined to each other with examples of pyramids, prisms,

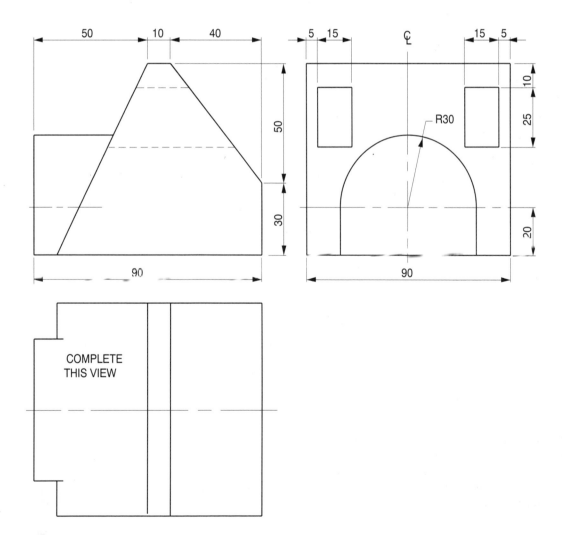

Figure 6.1

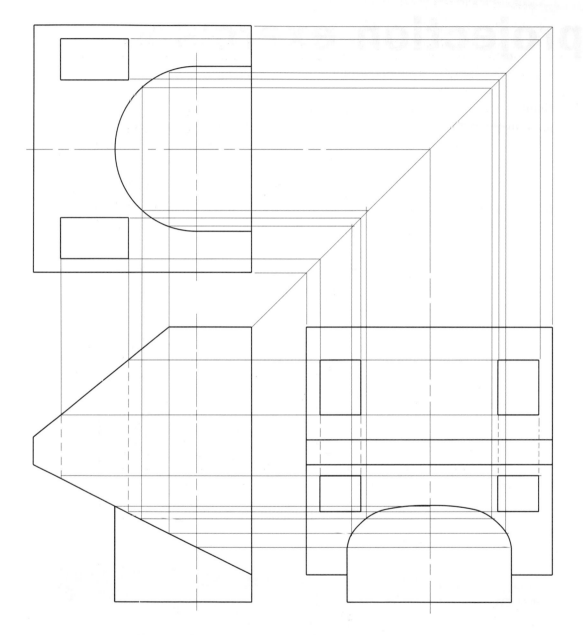

Figure 6.2

cones and sphere. Examples show profiles formed by the manufacturing processes of moulding and casting. Auxiliary projection examples illustrate cases where components are viewed at an angle.

Plotting boundary lines

The ability to be able to project details from one view to another in order completely to define three-dimensional objects on a two-dimensional surface is a very necessary part of draughtsmanship. The following exercises consist of a variety of shapes and forms and requires the draughtsman to transfer edges, corners, surfaces and points to complete views in orthographic projection.

Figure 6.1 shows a component drawing to be completed. Note the presence of a construction line at 45° from the corner of the front view. Note also the positioning of the other two views and remember the simple rule that widths in the end view are equal to depths in the plan view.

The professional draughtsman would not transfer and reposition details in this way by the use of a 45° line. With experience this becomes an automatic manipulation where you measure in one view and redraw in the other. However, I hope this increases

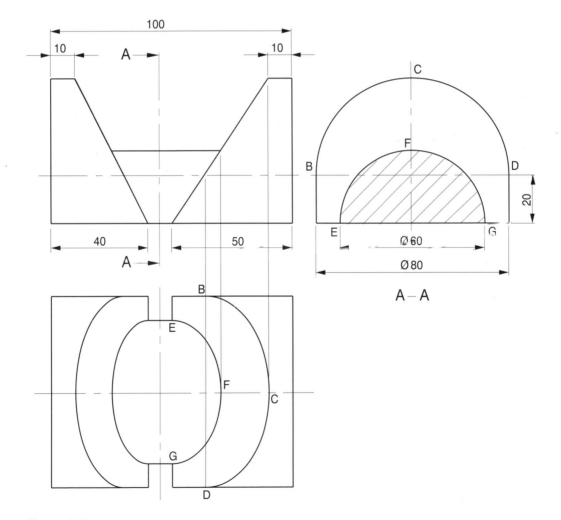

Figure 6.3

understanding now. Complete the plan view and add the hidden detail lines and dimensions. Erase the construction lines on your finished solution. Note that the curve in the plan view is a semi-ellipse.

The component illustrated in Fig. 6.3 consists basically of two semicircular parts which are cut by flat surfaces. The four curves in the plan view are semi-ellipses which can be added after positioning the major and minor axes. Construction lines are included to fix the necessary points BCD and EFG. Draw the ellipses and remove the unwanted parts. Finish the outline with a 0.4mm polyline and add the dimensions and section plane.

The example in Fig. 6.4 shows a tapered block with a curved surface, and it is required to project a plan view. In the front view, four random points have been taken along the curve. The width at each position is obtained by looking directly onto the face, and the measurement from the end view is transferred to the plan. Dimension P is typical.

A line through the intersections from the view above will give the required curve. Use the POLYLINE command and then the FIT option to modify the straight lines into a curve passing through the points of intersection.

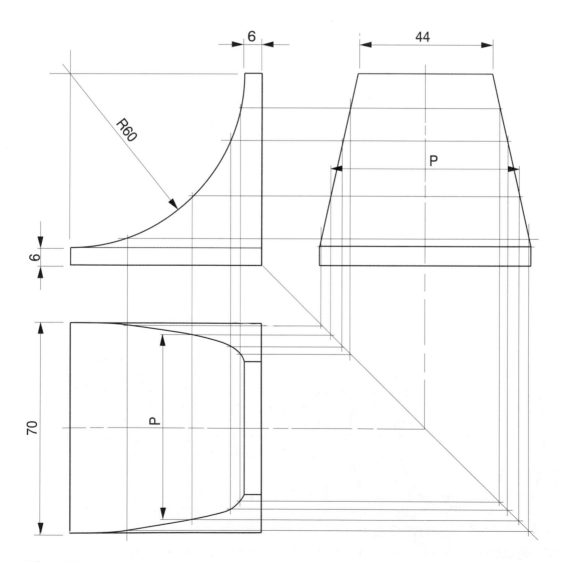

Figure 6.4

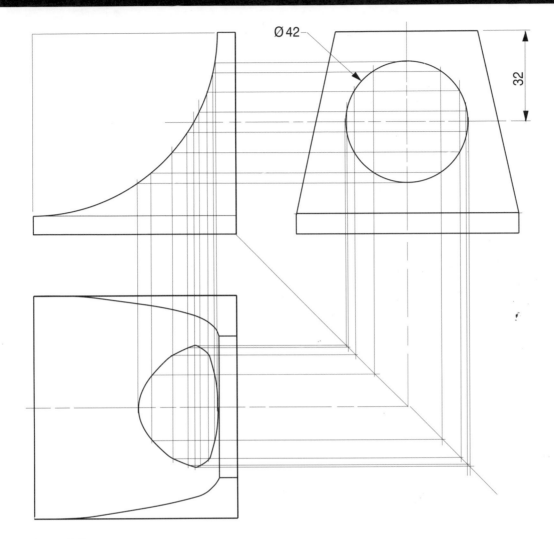

Figure 6.5

Figure 6.5 shows a similar method of construction and illustrates the plan of the profile on the curved surface due to the addition of a horizontally bored hole. Machining processes often give contours which need to be plotted from first principles by the draughtsman.

The projection exercise in Fig. 6.6 shows a hexagonal bar which is cut at two angles and with a vertical hole. Start the solution by drawing the hexagon around a 76 mm diameter circle with tangential lines at 60° to intersect at the six corners. Line in the outline as shown. Note that the width of the end view is equal to the depth in the plan view (76 mm). Project the front and end views and position the corner points. The projection of the hole from the top face in the front view will give an ellipse in the end view, and this can be drawn easily after projecting the positions of the major and minor axes.

The example in Fig. 6.7 shows two views of a cylinder which is inclined at an angle of 45°. The cylinder is cut horizontally by a plane passing through one corner. Draw a plan view and hatch the cut surface. Then add the hidden detail.

The following method should be used. There are two ellipses in the plan view. Draw the large one first, modify the outline to a polyline, then use the HATCH option in the

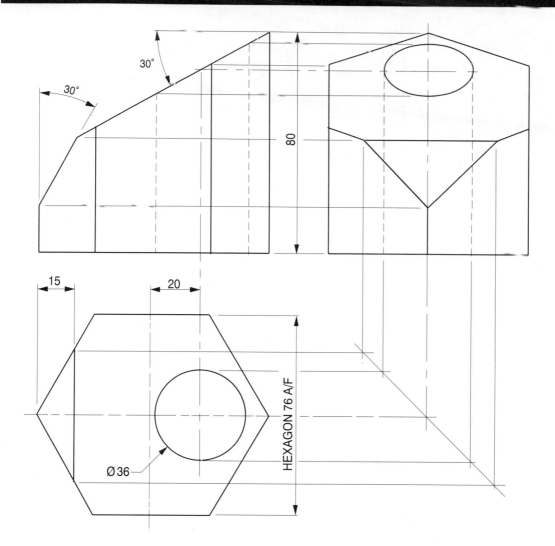

30°

30°

80

15

20

HEXAGON 76 A/F

Ø 36

Figure 6.6

Draw menu. Draw the ellipse for the bottom of the cylinder in projection and delete the right-hand half. Modify the left side to a polyline. On the right side of the plan view draw another ellipse of the same size using the dashed thin line. Use the BREAK option to delete the left hand side, then the MOVE command to join up with the thick line semi-ellipse.

The exercise in Fig. 6.8 shows a square prism positioned at an angle and cut by a horizontal plane. It is required to draw a sectional plan view. All the construction lines are included to copy the solution.

Projection from conics

The following three exercises illustrate important curves from a study of conic sections. The algebraic relationships are dealt with in considerable detail in mathematics courses.

Figure 6.9 shows a solid cone, and a vertical cut is taken parallel to the centre line. If a complete cone is cut parallel to its base, then the plan view will be a circle. In this case the

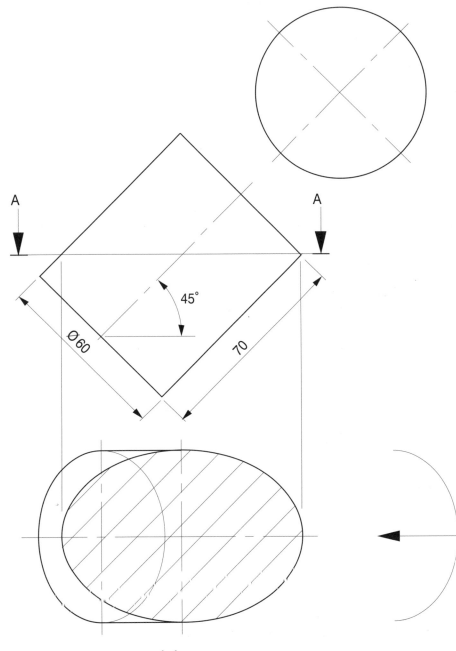

45°

Ø60

70

A

A

A-A

Figure 6.7

curve in the end view is obtained by drawing a section through the cone and plotting the position of points on the circumference where the cut face intersects with the circle. A typical section is shown to establish points A and B.

At point 4 the radius of the circle is distance XY. Draw this circle as shown; it intersects with the cut face on the plan at points A and B. Transfer width AB to the end view and repeat the procedure. Join the points to give a polyline and use the FIT feature in

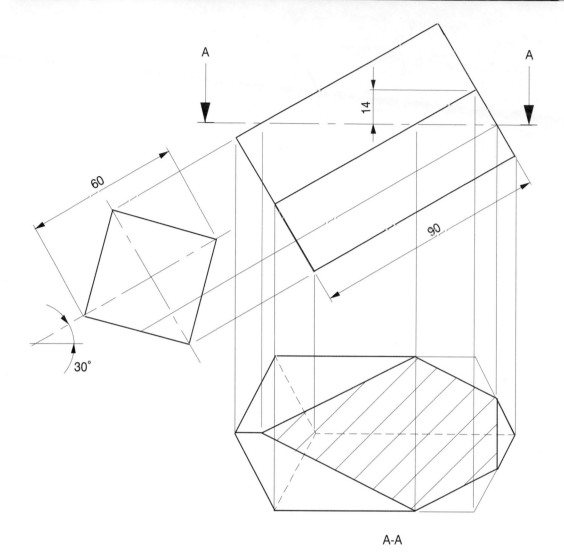

A-A

Figure 6.8

EDIT POLYLINE to produce the curve through the points. The distance between each section is left to the draughtsman's discretion. The curve in the end view is a rectangular hyperbola.

In Fig. 6.10 the cone is cut parallel to one side and the view of the cut face in the direction of arrow P gives a parabola. The procedure for drawing the curve is similar to the previous example.

The projection of the whole cone gives the auxiliary view shown in Fig. 6.11. Tangency points exist where the cut face crosses the centre line of the cone and a circle in each view will fix the positions on either side of the centre line. An 80 mm circle gives the major axis of the ellipse for the base of the cone in the auxiliary view. A line projected from the bottom right-hand corner of the cone establishes the minor axis, and the ellipse can thus be drawn.

Draw tangent lines touching the ellipse to complete the drawing. Use a large ZOOM value for accuracy. The tangent to the ellipse can be drawn using the OBJECT SNAP mode.

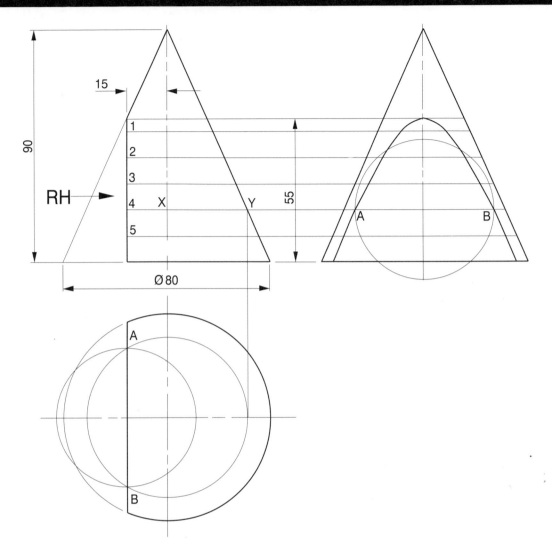

Figure 6.9

Figure 6.12 shows the dimension of a cut at 45° which touches both sides of the cone. The true face of any section plane passing through both sides of the cone is an ellipse. Ellipses are easily drawn in CAD, but you need to know the dimensions of the major and minor axes and the position of the centre point. In Fig. 6.12 bisect the line AB to give midpoint M. Draw line XY parallel to the base and a circle of diameter XY in the plan view. Project a vertical line through M to give points S and T. The minor axis of the ellipse in the plan view is distance ST. Use the ELLIPSE command to complete the plan view.

The remainder of the drawing is detailed in Fig. 6.13. Project width ST to the end view also points A and B and construct the ellipse. Project points E and F to the end view, and draw polylines to establish the outside limits of the cone.

The auxiliary view is constructed by drawing the ellipse projected from the base of the cone and the other ellipse projected from the cut face. The construction lines have been left on the solution. Two tangential lines are also drawn from the projection of the apex of the cone to the base ellipse. Line in the drawing as shown with the side polylines and hidden detail for the rear portion of the base.

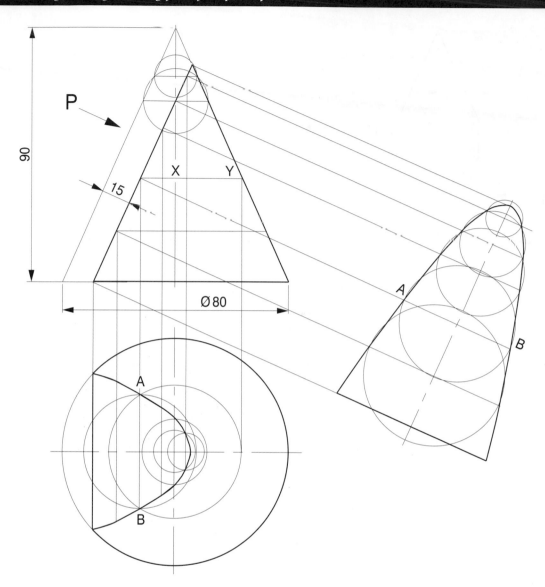

Figure 6.10

Interpenetration lines between solids

Intersection of square pyramid and prism

The example in Fig. 6.14 shows the outline construction for two intersecting solids. In order to draw the three views, it is essential to determine the lengths of all the lines and intersection points. If a model was to be manufactured in thin sheet metal, a pattern or template would be required, and the method is described later. Draw the outline on a suitable GRID and ensure that the centre lines are located using the SNAP feature. To complete the drawing it is necessary to fix the position of points P and Q where the corners of the prism touch the sloping faces of the pyramid. If the solid is cut across horizontally 30 mm above the base, then the bottom plan view will be seen and clearly a rectangular shape intersects the square at the required points. Project these points to the front and plan views to complete the drawing.

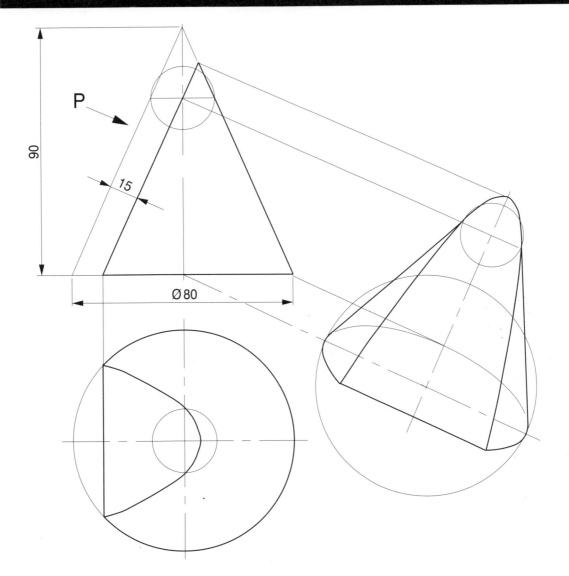

Figure 6.11

In the second example (Fig. 6.15) the square prism is inclined at an angle of 45° to the base. The construction lines for the polygon ABDEC are required to establish the points R and S.

The projection exercise in Fig. 6.16 shows a hexagonal pyramid with a triangular prism passing through its sides. Redraw the views shown, commencing with the plan and the hexagonal base. The dotted hexagon will give the profile parallel to the base and 25 mm above the base. All the relevant construction lines are given. It is necessary to establish all of the corner points in each view before drawing a pattern development of the two parts.

In Fig. 6.17 a cone and sphere intersect and the exercise shows the method of drawing the curve of interpenetration. Set out the cone and sphere to the given dimensions. With a relatively small screen I find it beneficial to line in those parts of the outline which I know will be needed in the final solution, as this procedure reduces the number of construction lines on the drawing.

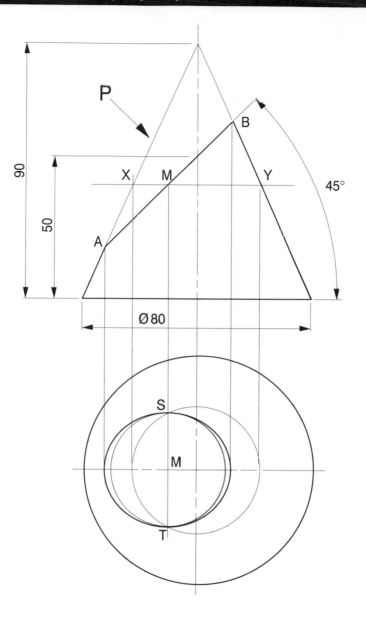

Figure 6.12

Transfer the points A and B, where the sphere intersects with the side of the cone, from the front view to the plan. PQ is any plane parallel to the cone base. If the assembly is cut along this plane and viewed from above, the profiles of the cone and sphere will appear as two intersecting circles, and in the plan view these intersections are C and D. Transfer a vertical line through these points to give point E on the plane PQ. Point E is a typical point on the required curve in the front view, and C and D are typical points on the plan view.

You may find that it is not easy to maintain accuracy if both views are on the screen at the same time, since the intersections such as C and D occur between circumferences which cross at oblique angles. This situation can be improved if the maximum ZOOM feature is used so that just the part of the plan with C and D appears on the screen. Draw a vertical line through them with the ORTHO feature on, to the top of the screen. Then use the EXTEND option to increase the line length to touch plane PQ. Points C, D and E can be

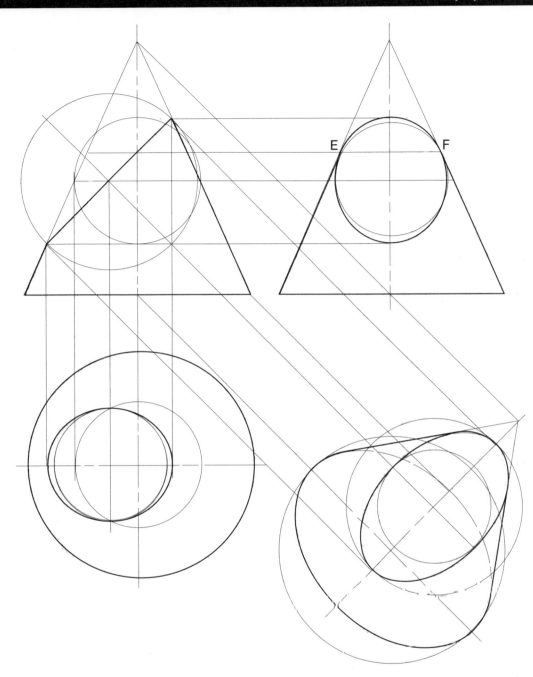

Figure 6.13

marked using the DOUGHNUT option, and a hole diameter of 0.5mm and outside diameter of 1.5mm are suitable. When the two intersecting circles are drawn, they can be exactly transferred if their centres are on SNAP positions.

Commence with a high ZOOM setting and the front view on screen. Use the **Circle** menu and choose **Centre, Radius**. Draw the circle from the centre line to the slope of the cone and follow this immediately with the same circle on the plan view, which will be accurate. Repeat the procedure for the other pair of circles. Having established points

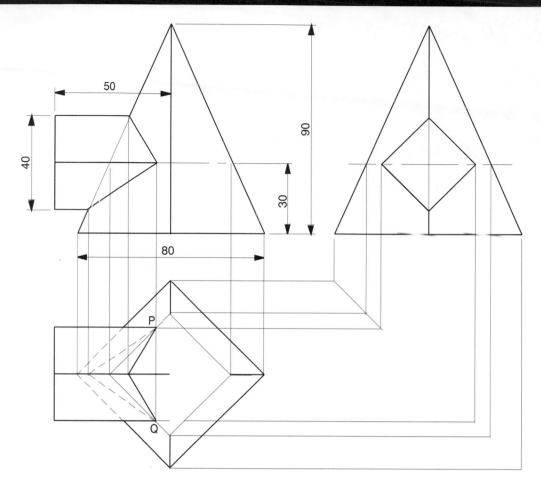

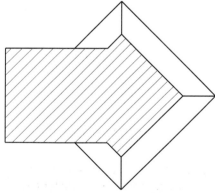

Figure 6.14

C, D and E, erase the construction circles to avoid unnecessary congestion on the screen with constructions on other planes.

Repeat the procedure for other vertical positions between points A and B to build up sufficient points for the curve. Line in the curve with a polyline and use the FIT option to give the best curve.

Since the cone and sphere lie in the same vertical plane, the two sides of the curve will be symmetrical, and it is inevitable that one side of the curve will be more accurate than the

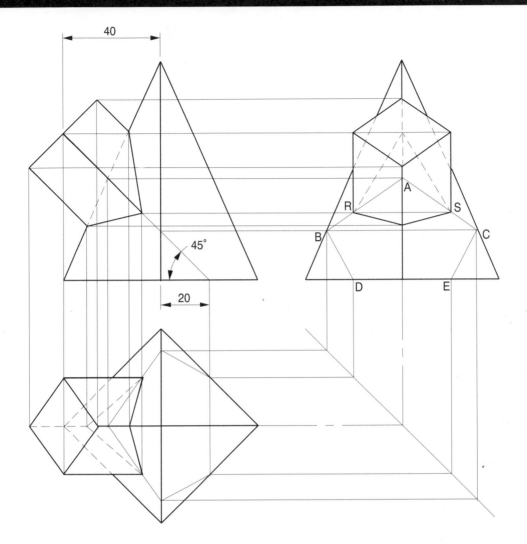

Figure 6.15

ther. You can if you wish then eliminate one half and use the MIRROR feature to reproduce two identical halves. Figure 6.18 shows the complete solution.

Many components are fabricated by joining pieces of standard raw materials such as plates, strips, rods, cubes, cylinders and prisms. Alternatively, shapes can be moulded into various forms. Manufacture using lathes, milling and drilling machines involves the removal of material and a change from the original shape.

The general appearance of a drawing for a finished product must include the edges, corners, joins and recognisable features after fabrication and machining have been completed. All of these aspects need to be projected accurately from view to view in order to provide useful and accurate drawings. It is often the case that lines need to be plotted, and this section shows various practical examples.

Figure 6.19 shows the dimensions of a moulded column base. The problem is to project the curves terminating at the four corners. It is only necessary to plot one corner line and use the MIRROR feature to copy about the vertical centre line, then repeat the procedure for the pair about the horizontal centre line.

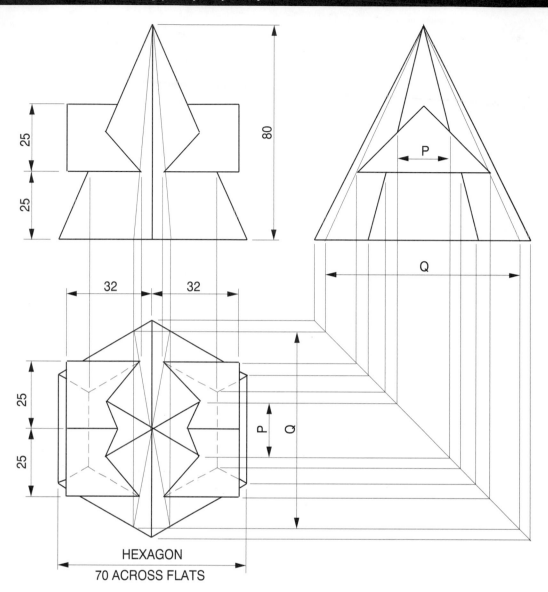

Figure 6.16

The solution in Fig. 6.20 illustrates the method. Draw a horizontal section line through the curves in the front view and project to the end view to give the points AB and CA. The width CA in the end view is equal in depth to CA in the plan view. The construction shows a line at 45° from the pillar corners and the projection lines to position points C and A in the plan view. Repeat the procedure with other section planes to provide sufficient points to give an accurate curve. Use the POLYLINE feature to connect the points and then the FIT option on the EDIT POLYLINE command to change the line into a curve.

In Fig. 6.21 the side of the column has been cut away and the modified front view is shown.

Two views of a casting are shown in Fig. 6.22. The rectangular form on the left is joined to the cylindrical part with 60 mm fillet radii at the sides and 20 mm fillet radii at the top. The

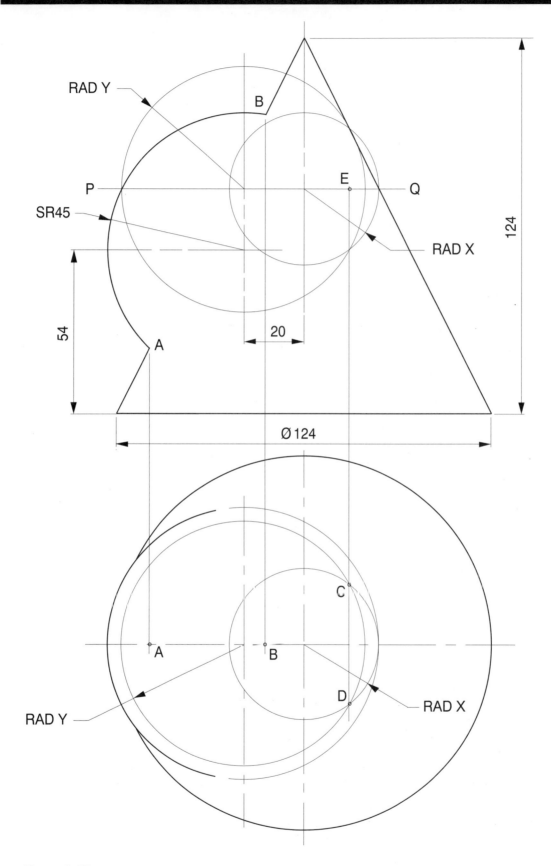

Figure 6.17

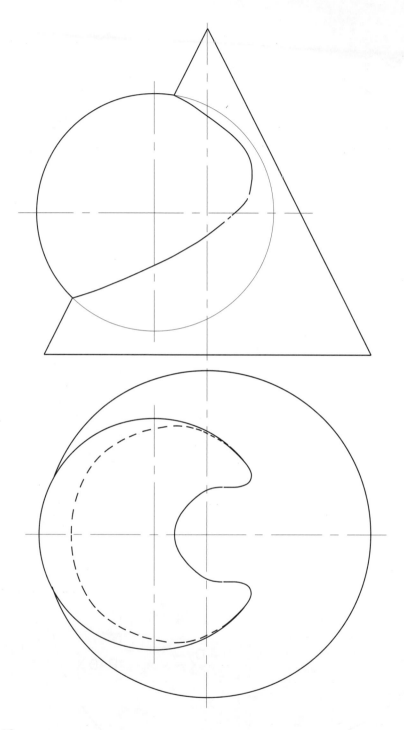

Figure 6.18

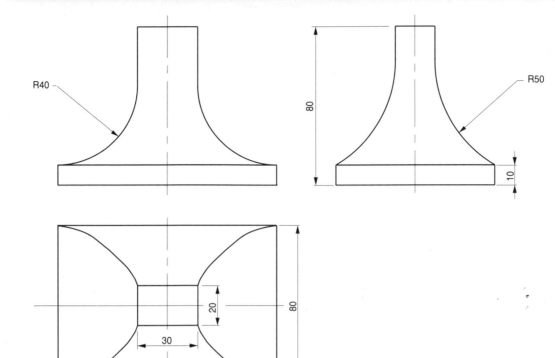

Figure 6.19

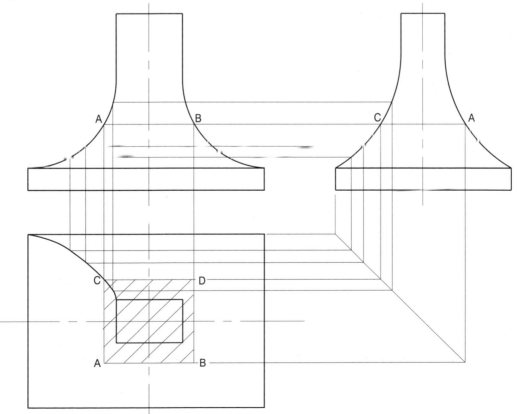

Figure 6.20

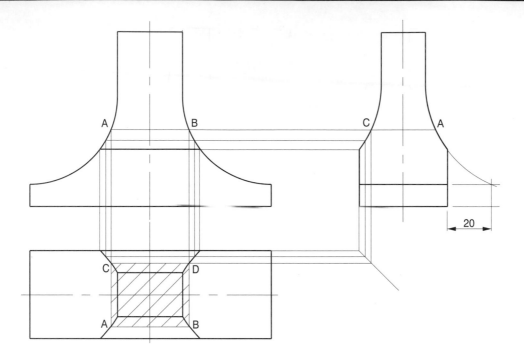

Figure 6.21

drawing would be incomplete without the curve between points A and E in the front view, which needs to be plotted. If the FILLET feature is used then centre G in the plan must be fixed.

Select **Line** in the **Draw** menu and take centre O as the first point, then select **Object Snap** in the **Assist** menu. Choose **Centre**, enter **OK**, pick anywhere on the fillet radii and the line will snap to centre G. Project lines from F and E to the front view.

B is any point on the top fillet curve. Draw a circle in the plan view with radius CB, then a vertical line from the point of intersection on the outside curve to give point B1. Repeat the procedure for other points as required. Draw a polyline through the intersections from A to E, then use the **Edit Polyline** feature in the **Modify** menu and select **Fit** to give the best curve.

A plastic handle grip for a gardening implement is shown in Fig. 6.23. The cylindrical form is designed with a hemispherical shape at the left-hand end and a domed finish on the right. A metal tube is pushed into the grip along the axial centre line. The flutes will leave irregular curves on the surface where they run out at each end. Above the centre line is the construction for the top left side, and below it for the bottom right-hand end. Plot the curved portions at each end then use the MIRROR feature twice to duplicate the curves. Join each pair with parallel lines projected from the corners of the flutes in the end view. Commence by drawing a circle tangential to the bottom of the flutes and project this diameter to each end of the grip. This determines the position of the end of the curves at each side. At the right side the dome leaves a corner line as shown but the spherical finish will not. These features establish the length of the curves at each end.

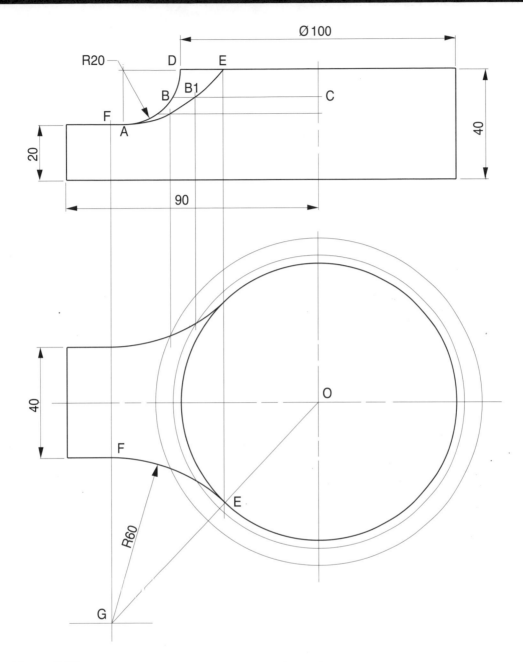

Figure 6.22

Start at the left side and draw a pair of vertical lines between the curve limits, and where they intersect the outside profile project circles across to intersect the flute profiles. Transfer the intersection points on the flutes back to the vertical lines but use the ZOOM command at its highest magnification for accuracy. It is also advantageous with a small screen to use the EXTEND option. The view with the flutes could take up most of the screen, but if lines from the intersections are accurately positioned and projected to the left side of the screen they can easily be extended. Use ZOOM again to draw a polyline through the left-hand plotted points. Now use the FIT option from the EDIT POLYLINE menu to convert the polyline into a curve. In Fig. 6.23 I have only left the construction

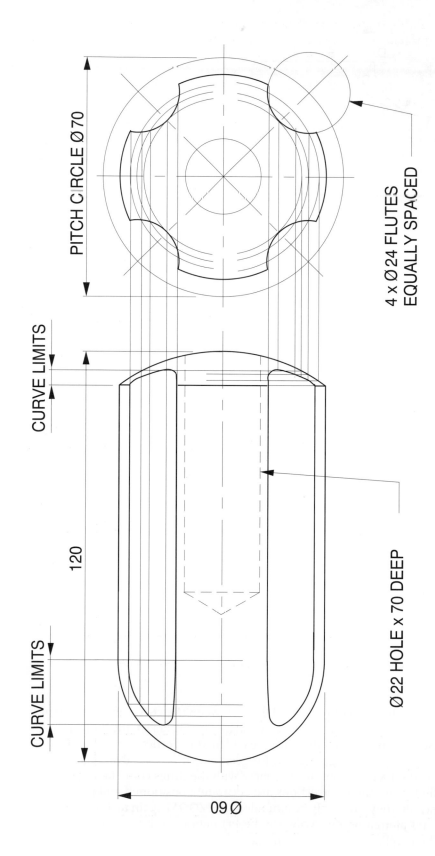

PITCH CIRCLE Ø70

4 × Ø24 FLUTES
EQUALLY SPACED

CURVE LIMITS

120

Ø22 HOLE × 70 DEEP

CURVE LIMITS

Ø60

Figure 6.23

lines above the horizontal centre line in order to make space for a similar application to draw the bottom right hand curve beneath the centre line. Accuracy is all-important, so use the ORTHO and SNAP buttons at every opportunity.

Auxiliary projections

Earlier orthographic examples were represented on the drawing sheet by projecting two or more views at 90° to each other. A draughtsman will try to orient a component to give the maximum amount of information to define completely all the component features, and this consideration determines the number of principal views on the drawing. However, some details may be positioned at an angle and additional views are necessary for clarity. Sometimes, new views of objects are required to be drawn at angles of less than 90°.

Figure 6.24 shows a small solid model of a church and it is required to project a new view from the given plan at an angle of 30°. The solution is given in Fig. 6.25 with the construction lines. Project lines from each corner and insert a new ground line A1–B1 at 90° to the line of sight. Dimensions from this new line will be the same as from the original ground line AB. Work from left to right on the plan view and line in the three main features.

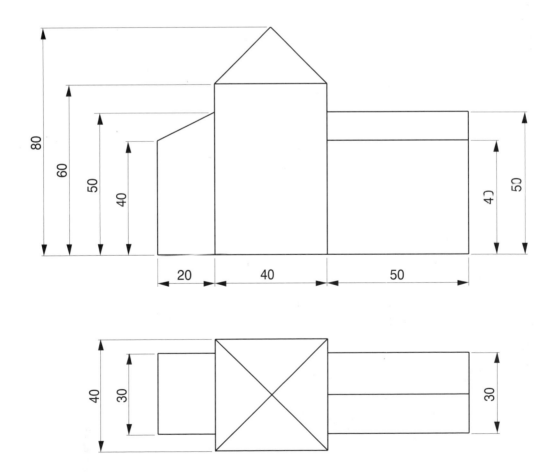

Figure 6.24

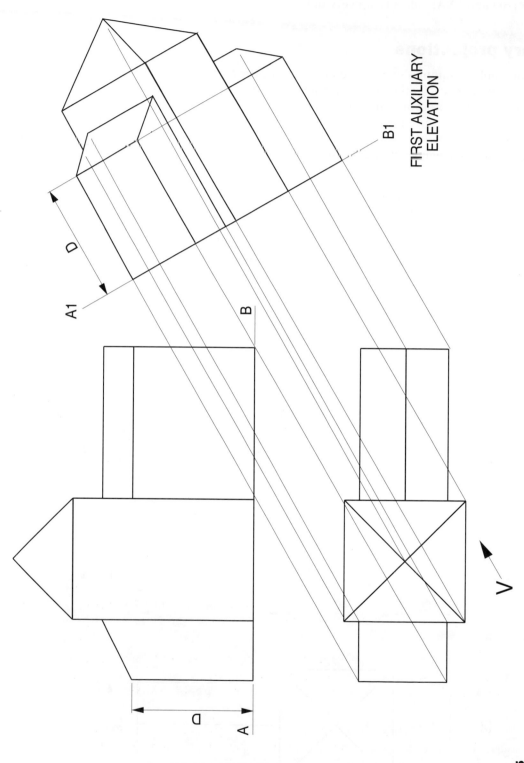

FIRST AUXILIARY
ELEVATION

A1

D

B1

B

D

A

V

Figure 6.25

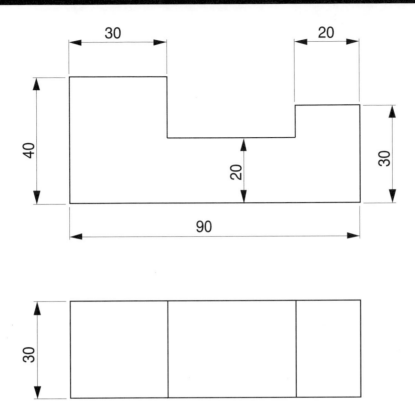

Figure 6.26

The next problem deals with the situation where, having drawn an auxiliary view, a second auxiliary view is required to be projected from it. It is helpful to use a model here to visualise the situation, since you can easily reposition it and look from different directions.

Figure 6.26 shows a small block cut from polystyrene packing with a knife. These dimensions allow a full-size solution to be positioned on the CAD screen. The solution is shown on Fig. 6.27.

Draw a first auxiliary view from the given plan view. The line of sight is at 45°. The block rests on a flat surface shown in the given front view as ground line AB. Position a new ground line A1–B1 at 45° and project construction lines from the corner points 1 to 8. The distances above the new ground line to each corner can be taken from the existing front view. Only three of the outside vertical corners of the block are visible. Line in the finished polylines as shown.

Take a small piece of card to act as a model for the new ground line CD and position it at point 2 at 90° to the line of sight. Pick up the model and card and rotate them through 90° in the direction of the arrow. The first auxiliary view should now be clear.

The second auxiliary view is drawn by turning the first auxiliary view through 90° vertically, still maintaining its angular position with respect to the new ground line.

On the screen drawing, vertical projection lines are taken from points 1 to 8. The dimensions from the new ground line C1–D1 will be the same as from CD to each corner on the original plan.

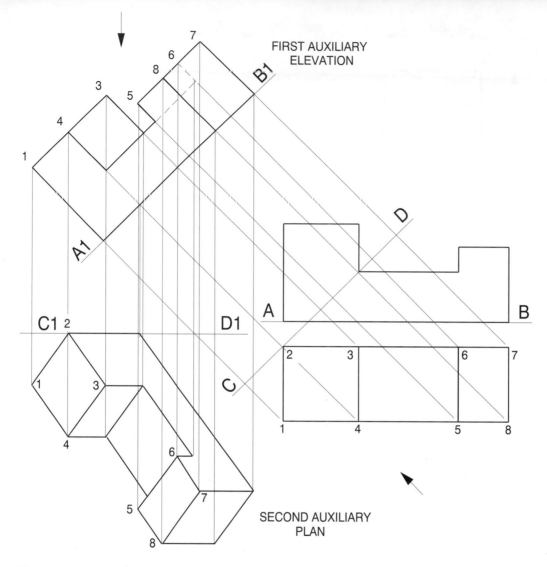

FIRST AUXILIARY
ELEVATION

SECOND AUXILIARY
PLAN

Figure 6.27

It is sometimes the case that solving projection problems on the small screen can result in a mass of confusing construction lines. The best advice here is to draw small units one at a time and remove unnecessary construction lines after the unit is complete. For example, plot face 5, 6, 7 and 8 and note that opposite sides are parallel to each other. Having satisfied yourself that this part is accurate, clean up the screen and continue with face 1, 2, 3, and 4. It is often a useful constructional exercise to take a preliminary print on which you can check, mark up and measure dimensions.

Figure 6.28 shows a cone which has been cut parallel to one side. The auxiliary elevation is complete and the object of the exercise is to draw the front view and the plan. The profile of the cut face viewed at right angles in line with arrow X will give a parabola.

Commence the solution by drawing the outlines of the cones. Draw line AA on the auxiliary view and a circle of diameter AA on the plan. Project lines up to the front

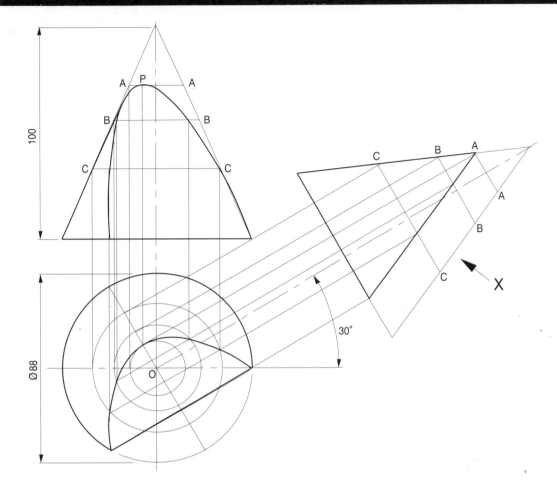

Figure 6.28

view to transfer line AA and fix the top of the cut face at point P. Repeat the procedure for points BB and CC, which are intermediate lines, to establish further points for the required curve. These points lie on the circumferences of the circles which interesect the cut face.

Chapter 7

Pattern development

By the end of this chapter you will be able to draw patterns for various hollow objects and make useful models to check dimensions, shape, appearance and the position of joins. These practical exercises assist in the overall comprehension of three-dimensional forms. Applications of this type of work are found in thin sheet metal and plastics fabrications. Development exercises include:

- Parts of rectangular and triangular prisms.
- Cylinders with branches.
- Cones and intersections.
- Hexagonal pyramid developments.
- True length applications.

Parallel line

This section introduces methods of drawing patterns from thin sheet material. If a print is then taken and the pattern cut out, it can be folded to make the component. One of the more difficult aspects of projection is trying to visualise a three-dimensional object on a two-dimensional piece of paper. Models are ideal in this respect for comprehension purposes, and also to appreciate the problem of true lengths where lines are placed at angles away from the plane of the drawing sheet. The exercises here are examples from rectangular, cylindrical, and conical applications.

Fig. 7.1 shows a component to be made from thin sheet material. We are required to draw a pattern. The left- and right-hand sides in the solution meet along the vertical centre line at the rear, while the bottom remains open. No allowances are made for the joins. It is only necessary to draw half of the pattern and then use the MIRROR feature to complete the drawing. Print a copy of the solution and cut out the pattern. The dotted lines show the positions of the bend lines.

In Fig. 7.2 the sides of the component are angled. Note that the lengths AB, BC, CD and DE represent distances along the centre line of the pattern. The widths CD and FG are taken from the plan view and should be drawn on the pattern to determine the true lengths of the corners. Print a copy and again make a model. True lengths are sometimes very difficult to visualise; measure lengths on the model to check.

Patterns are needed to manufacture the component shown on Fig. 7.3, which consists of parts of a triangular prism and a cylinder. The pattern is required in three separate parts, namely the top and bottom surfaces and a continuous strip for the cylindrical section between the two end wings.

Draw the given profile and rotate the component anticlockwise through 30°. Divide the plan view into 30° segments and transfer the points on the circumference to the front view. The back and front profiles of the cylinder are similar.

Divide half of the circumferential length of 36π into six strips of 18.85 width using the RECTANGULAR ARRAY feature. Transfer the heights from the front view as shown and draw a polyline through each. Use FIT on the MODIFY POLYLINE option to complete the

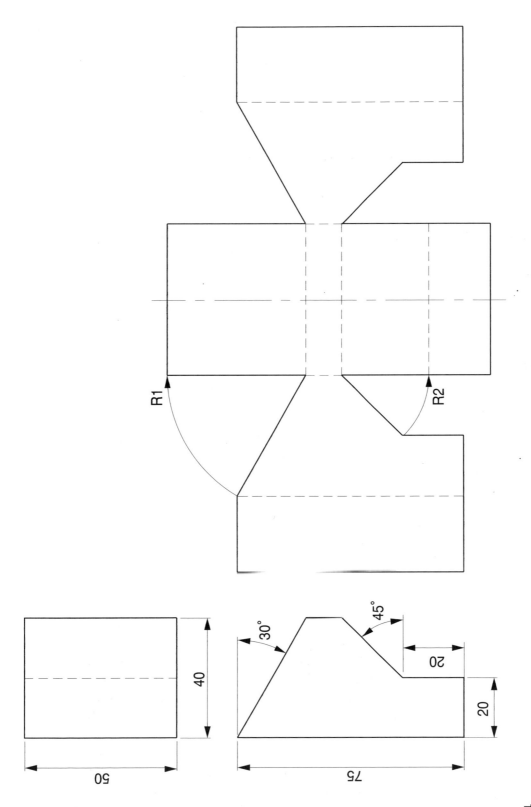

Figure 7.1

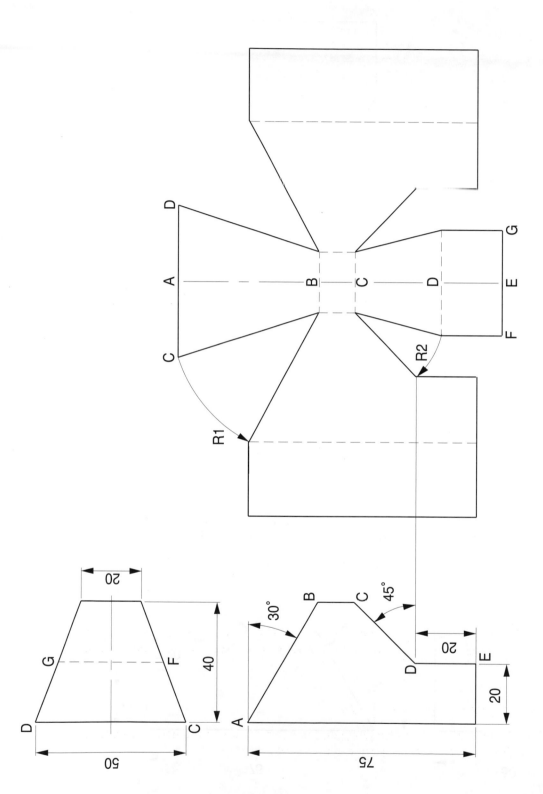

Figure 7.2

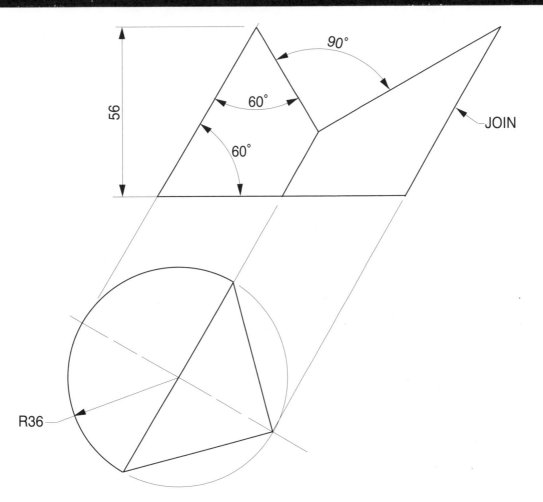

Figure 7.3

curves (Fig. 7.4). Note that the vertical heights from the side wings can be transferred directly but the true widths of each wing must be taken from the plan view.

The pattern for the top can be constructed from a semi-ellipse and the construction lines show the method for finding half the major axis OA (Fig. 7.5). The construction is similar to give the centre line distance OB for the triangle. The bottom surface is constructed in a similar manner.

Take a print and cut out the patterns to make a model. Particularly note the true lengths of lines that lie at an angle to the surface of the paper. The dotted lines indicate bend lines.

True length exercise (Fig. 7.6)

This small example is included to draw added attention to the true lengths of lines at compound angles. Construct the pattern and cut it out to study.

Cylindrical pattern development

A pattern is required for an elbow fitting detailed on Fig. 7.7. The two halves are identical in shape, but when manufactured they can be cut from metal strip more economically if

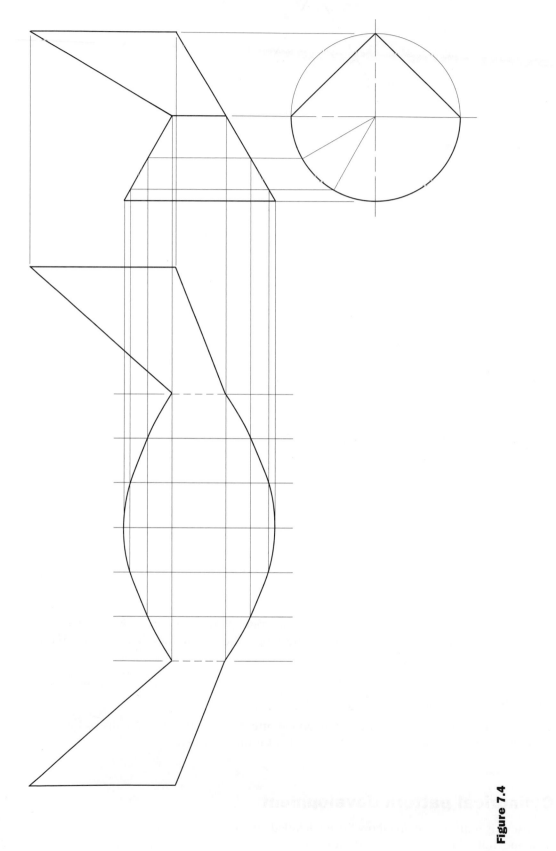

Figure 7.4

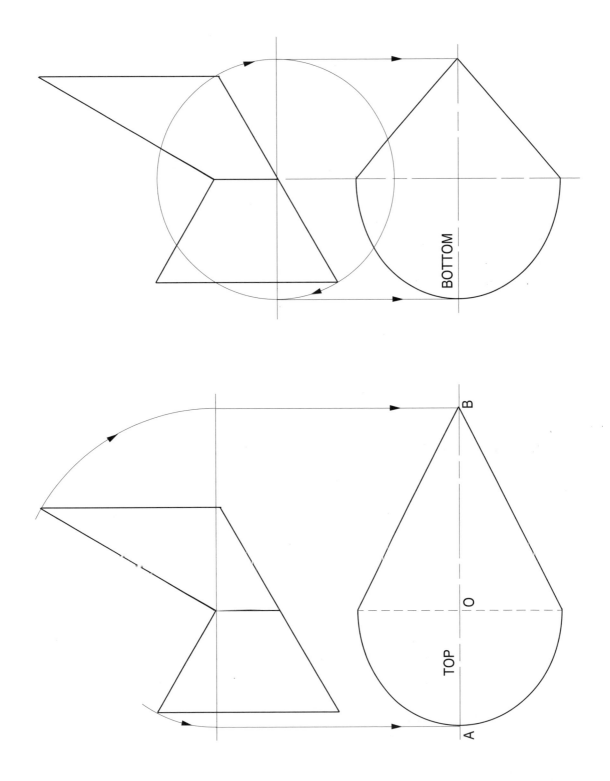

Figure 7.5

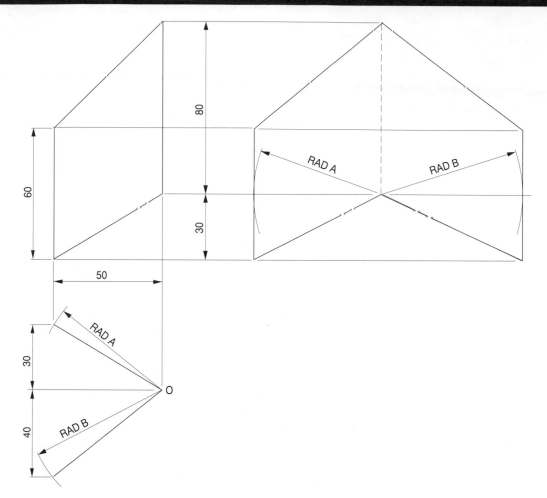

Figure 7.6

the join lines at the sides are staggered by 180°. Join lines are generally arranged along the shortest sides. However, the two options are shown below.

The circle can be divided into 12 equal parts using the ARRAY feature. Draw line O1. Choose POLAR ARRAY, click on the circle centre and insert 12 items. Then add the numbers as shown. The pattern will be 40 π in length, which is 125.66 mm, and the distances around the circumference between successive points are calculated to be 10.47 mm.

The vertical lines on the pattern can be set out using the RECTANGULAR ARRAY feature. My software requires me to perform these operations:

1. draw a vertical line at point 1 which is to be copied;
2. choose **Array** in the **Construct** menu;
3. choose **R** for rectangular array;
4. type **1** for the number of rows;
5. type **13** for the number of columns;
6. type **10.47** which is the distance between columns;
7. press <Enter>.

Please note that other software will divide a quoted distance into a given number of parts directly.

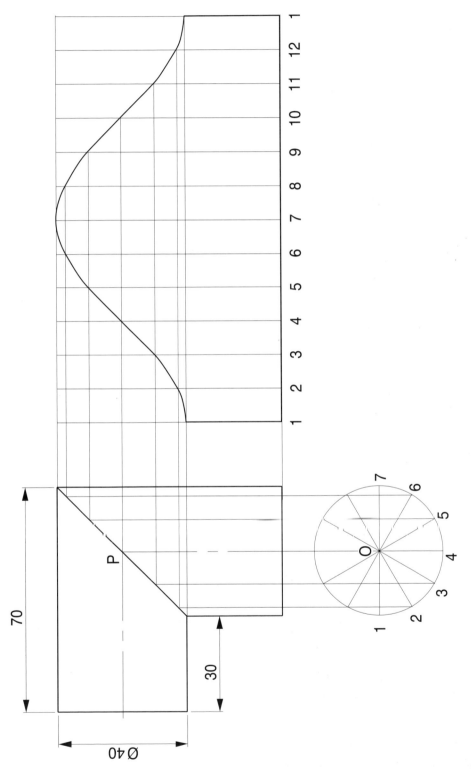

Figure 7.7

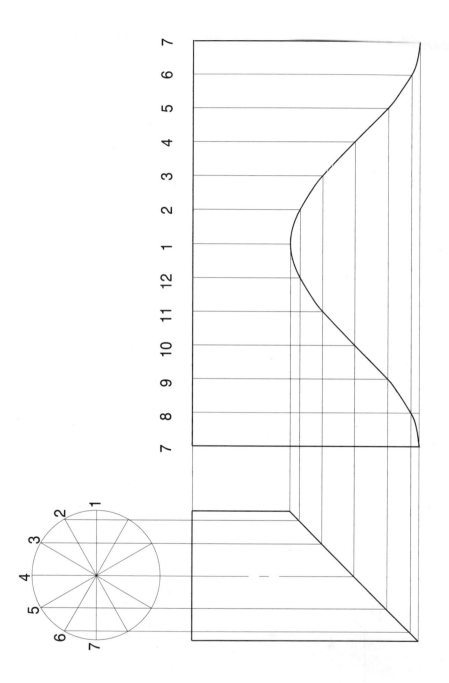

Figure 7.8

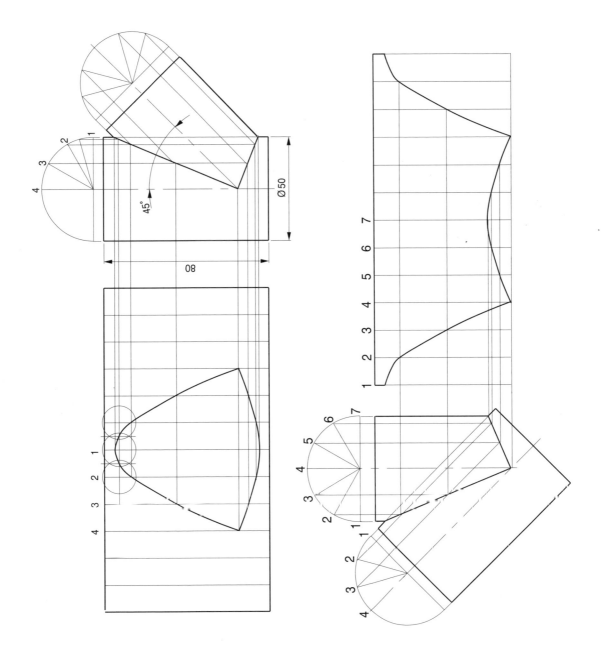

Figure 7.9

Transfer lines from the intersections on the sloping face to the pattern as shown. Draw a polyline through these plotted points. Select **Fit** from the **Edit Polyline** feature to obtain the curve.

It is an interesting AutoCAD exercise to draw the illustration in Fig. 7.8 using details from Fig. 7.7. Save your drawing with a new file name and erase the dimensioned part on the left. Rotate the elevation through 180° about centre P. The pattern profile is already in position but it is necessary to make modifications to its size and the numbering. Confirm the results by taking a print from each drawing and cutting out the patterns to make a model.

Cylindrical branch

The dimensions of a cylindrical branch are shown on Fig. 7.9. Provided the main duct and the branch are the same diameter, then the intersection between the two on the elevation will be formed from straight lines.

The main duct requires a hole to be cut to insert the branch. The procedure for the pattern length and its division is the same as that previously described. To improve accuracy, an intermediate point can be added between points 1 and 2. Use the ZOOM feature and draw a small line at 15°. Project a vertical line to the intersection, and as much of the horizontal line from the intersection to the left that the screen area will allow. Use the EXTEND option to extend this line to the pattern. To fix the position on the pattern, bisect the distance between points 1 and 2 by drawing small intersecting circles of the same radii and drawing a line through their intersections.

The pattern development for the branch pipe is inconvenient to draw if projected at an angle from the given view. A better method is to make a copy of the branch and rotate it through 45° anticlockwise to bring the required part into the vertical position. The complete pattern is shown so take a print of your solution and make a model.

Development of hexagonal pyramid

The dimensions of a hexagonal pyramid are given in Fig. 7.10 and the pattern construction for the surface on Fig. 7.11.

Number the corners and commence the pattern by drawing a radius equal in length to the slant height O4 (Fig. 7.11). Draw line O2 at a convenient position. Draw an arc of radius R2 on the plan from point 2 on the pattern circumference to give point 1. The complete pyramid consists of six triangles similar to O12. Use the ARRAY feature to draw and number them as shown.

The pyramid is cut vertically in the plan view and the surface 7,8,9,10 needs to be positioned on the pattern. Extend the line from 8 to 9 in the front view to the side of the pyramid at point P and draw radius OP to intersect lines O5 and O6 on the pattern. This construction is necessary to determine the true length of distances O8 and O9. Draw an arc of radius R from points 4 to 10 in the plan from centres 1 and 4 on the pattern in order to fix points 7 and 10 on the pattern. Join the points as shown to complete the pattern development.

Conical development

Part of a cone made from thin sheet metal is shown in Fig. 7.12. A pattern is required for the curved surface; it is cut from part of a circle whose radius is equal in length to O1, the

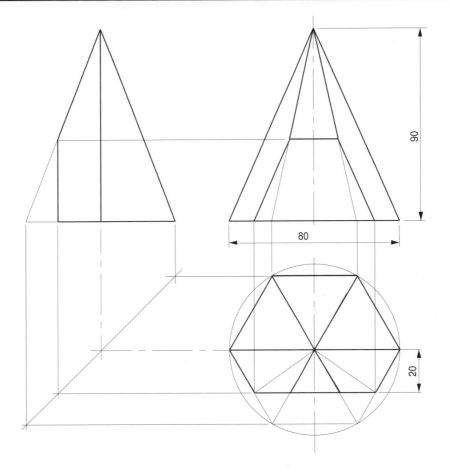

Figure 7.10

slant height of the cone. The length of the circumference of the cone base is marked off around the arc of radius O1 after calculating the angle 'A' :

$$\text{Pattern angle} = \frac{2 \times 35 \times \pi}{2 \times O1 \times \pi} \times 360 = \frac{35}{69.46} \times 360 = 181.14$$

Divide the base of the cone into six equal parts. This can be done by drawing radial lines 30° apart. However, try using the ARRAY feature and start this construction by drawing line B1. The following sequence will then be applicable. Choose **Array** from the **Construct** menu and the command line reads

```
Command: _array
Select objects
```

Click on line B1 and the command line reads `Rectangular or Polar array (R/P)<R>`. Type **P** and the line changes to `Centre point of array`. Click on point B the circle centre and the command line now reads `Number of Items:`. Type **7** which includes the first and last lines, and the command line reads `Angle to fill (+=ccw, -=cw) <360>:`. Type **−180** and the line reads `Rotate objects as they are copied <Y>`. Press <Enter> and the radial lines are drawn accurately.

Number the points 1 to 7 around the semicircle circumference and add the vertical lines to the base of the cone. Draw the slanting lines from the base to point O, the apex of the cone. It is necessary to determine the true lengths of these slanting lines from the base to

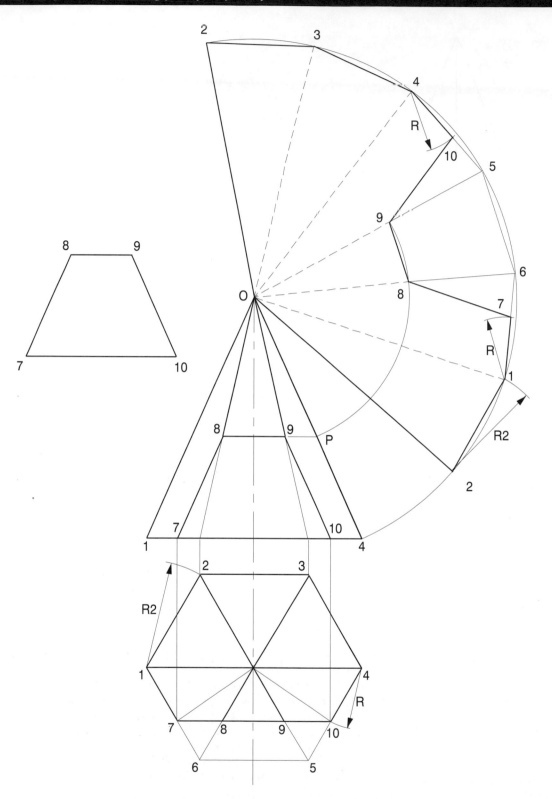

Figure 7.11

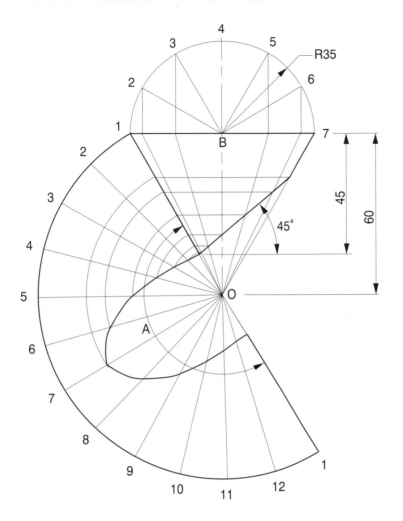

Figure 7.12

their intersections with the sloping face on the pattern in order to determine the shape of the profile. True lengths are obtained by drawing horizontal lines from each intersection to line O1.

In order to draw the part of the circle for the pattern, insert an arc of radius O1 and divide it up accurately by repeating the above procedure. Select line O1 and continue as before, but insert 13 for the number of items and the angle to fill will be 181.14°.

Number the lines around the circumference and draw the arcs from the true-length construction points on line O1. Draw a polyline through the intersections on the pattern for points 1 to 7, then use the **Edit Polyline** option in the **Modify** menu and choose **Fit** to give the best curve. The MIRROR feature about line O7 gives the other half of the pattern for lines 8 to 1.

Take a print and cut out the pattern to check the construction and projections.

Jug

Fig. 7.13 shows the dimensions of a small jug manufactured by cutting two conical patterns and soldering them together. The solution is in two parts in order to show clearly

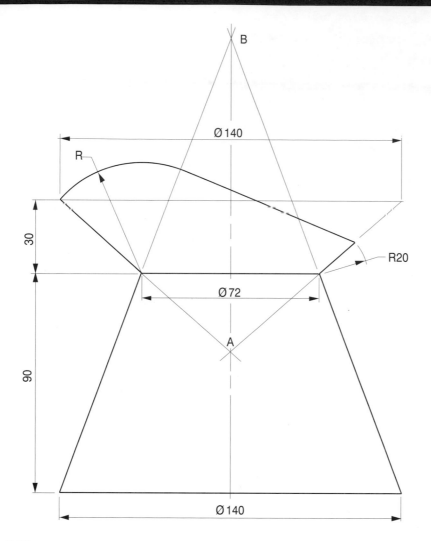

Figure 7.13

the construction lines. Draw a horizontal line tangential to the top of the jug and extend the two tapered sides in both directions to give the points 1, 13, and O (Fig. 7.14). Draw the semicircle. Use the length B1 and the ARRAY command to give the radial lines 1 to 13. Remember to insert 14 for the number of lines in the polar array.

The method I adopt for spacing the numbers 1 to 14 equally around the circumference is to draw a second circle parallel to the first, acting as a centre line, then use the TEXT command to place the numbers in position. Any final adjustment necessary can easily be made with the MOVE feature. Alternatively, if the numbers are needed several times, I make a list in a convenient space away from the drawing and then reposition them with the COPY feature. Erase the construction circle later when you are satisfied with the result.

Draw the vertical lines from points 2 to 12 to the base of the cone as in the previous example. Add the radial lines to intersect with point O. You will find that not all of the vertical lines touch the line 1 to 13 in 'snap' positions. For each line commence at point O with SNAP engaged and draw the line passing through the required intersections to a 'snap' point above, then use the TRIM command to tidy up the linework. Alternatively, for every line turn the SNAP feature off to position the lines accurately on the

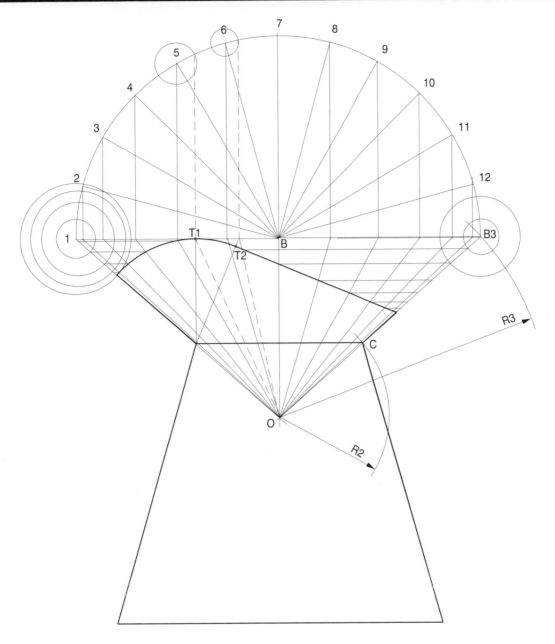

Figure 7.14

intersections on line 1 to 13. Ensure that you use the maximum ZOOM setting to give the best results.

For the pattern construction, copy radii R2 and R3 from centre P, equal in length to distances OC and O13. Draw one radial line P1 on the pattern and use the method described in the previous example to divide the pattern angle into 24 divisions. The total number of items will be 25 (Fig. 7.15).

There are two particular points which should be added to the pattern drawing for accuracy, and these are shown as T1 and T2. You will need to transfer two extra radial lines on the pattern drawing. Draw a dotted line from centre O to point T1, which is the tangency point between the curve and line B1. The second dotted line is taken from centre O through T2 to line B1. Point T2 is the tangency point between the straight line

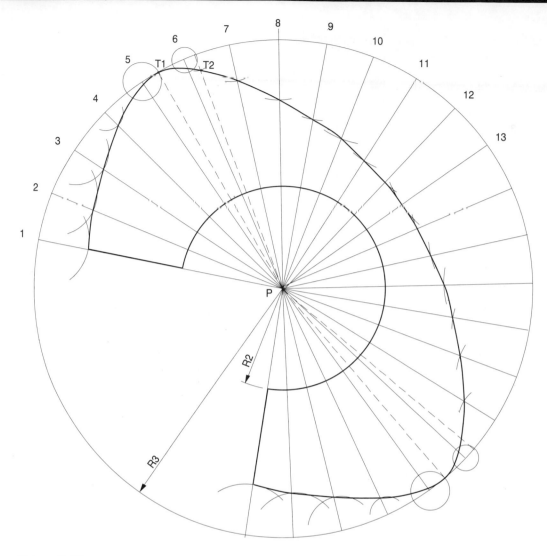

Figure 7.15

and the curve. Project vertical dotted lines to the circumference and add the two dotted circles from centres 5 and 6. Copy these circles on the pattern drawing and add the radial lines. Point T1 lies on the outside circumference, but a true length distance from point 13 is needed to fix T2.

I find it convenient to transfer distances accurately by drawing circles one by one with a large ZOOM value. The CIRCLE feature measures the radius and permits the operator to draw the next circle by clicking on the appropriate centre. Some construction circles are left here to emphasise the method, and these are erased afterwards.

Draw the polyline as before through the plotted points and use the MODIFY feature, choosing FIT to give the best curve. Use the MIRROR facility to complete the construction.

Fig. 7.16 shows a square prism which is fitted into a cone. If the prism is assembled from above, it will first enter the cone at section AA, making contact at the centre line. The lowest point will be reached when the corners touch the cone at section EE. The curve of interpenetration is plotted here by dividing the face of the prism into 10 mm strips and projecting them to the plan view. Draw radial lines O1, O2, O3, O4 and O5. Project radii

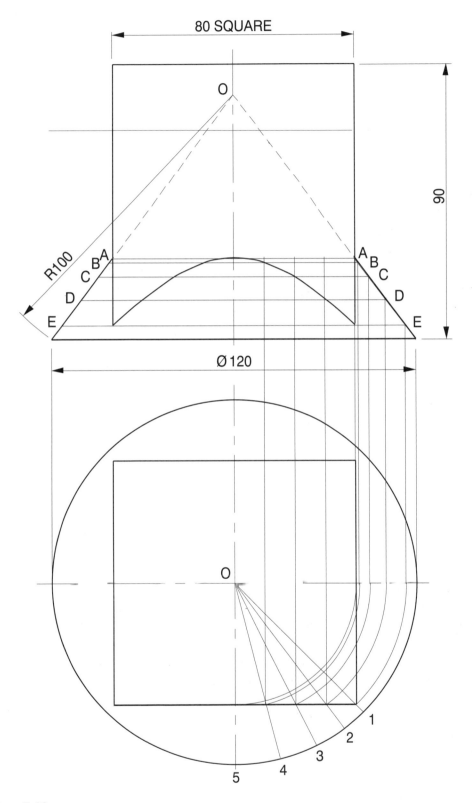

80 SQUARE

R100

Ø 120

90

O

A B C D E

Figure 7.16

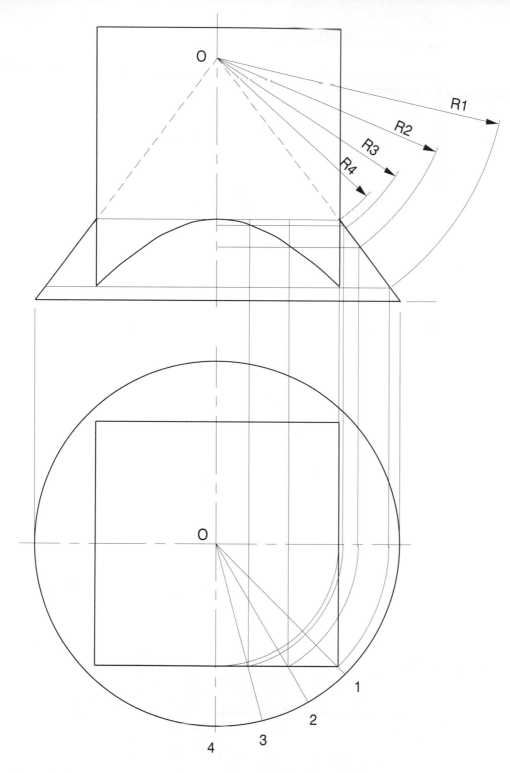

Figure 7.17

from the centre O through the intersections with the front face to the centre line in the plan and vertically up to intersect the sloping side of the cone. Add the horizontal section lines BB, CC and DD. Draw a polyline through the intersections as shown and use **Fit** on the **Edit Polyline** option from the **Modify** menu to complete the curve. Note that the

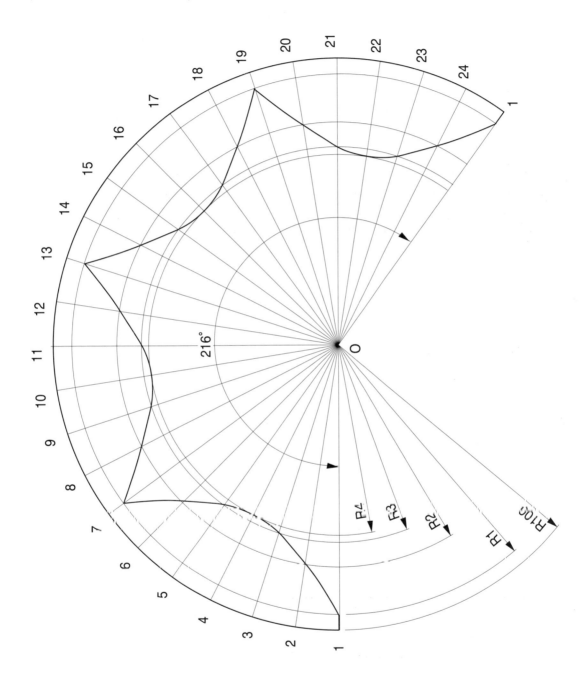

Figure 7.18

distances between points 1, 2, 3, 4 and 5 are not equal, but if the curve is only required to complete an assembly drawing then this construction will be satisfactory.

Now draw a pattern development for the cone by saving this drawing under a new file number and deleting all construction lines. Figure 7.17 shows the new constructions where the angles between O1, O2, O3 and O4 are equal and 15°. Project radii from the plan to the sloping side in the front view and establish the true lengths from the top of the cone to the intersections equal to R1, R2, R3 and R4. Figure 7.18 shows the constructions for the cone pattern. Draw part of a circle of radius R100, the slant height of the cone. The angle of the pattern is

$$\frac{120 \times \pi}{200 \times \pi} \times 360 = 216°$$

Draw the line O1 and plot the pattern blank by the method shown earlier for Fig. 7.12 using the ARRAY feature. It is only necessary to draw a quarter of the pattern curve using the true lengths as shown between points 1 to 7. Repeat the profile a further three times to complete the construction. The MIRROR feature can be used but ensure that you start the construction with vertex O on a GRID point for accuracy in defining the mirror lines. Alternatively, you can use the ARRAY feature.

Chapter 8

Fastenings

This chapter introduces commonly used fastenings which are necessary for assembly drawings. It is possible with CAD to store frequently used details in your own data bank, then recall and reposition items, as required.

At the end of the chapter you will be able to draw accurately:

- Nuts, bolts, studs and washers from standard dimensions.
- Understand the terminology and conventions relating to nuts and bolts.
- Draw fastenings on sectional views and assemblies.
- Draw isometric views of nuts and bolts.

Nuts and bolts

These exercises show accurate projections of nuts and bolts drawn from first principles and using standard information. Parts of these drawings can be used in other assemblies. They can, if desired, also be stored in block form in your own database to build a library of regularly used components to be used on assembly drawings.

ISO metric precision hexagon bolts, screws and nuts are covered by BS 3692 and ISO 272. The Standard also includes washer details. Tables of dimensions are listed relating to the thread size. An M36 nut fits an M36 bolt and the 36mm dimension is the outside diameter of the bolt. A large bolt of this nature would be used in structural steelwork assemblies and is an ideal size for this example. The dimensions are as follows (Fig. 8.1):

nominal thread size diameter 36
minor diameter of thread 31
width across corners (A/C) 62.5
width across flats (A/F) 55
height of bolt head 23
thickness of normal nut 29
thickness of thin nut (locknut) 14
washer outside diameter 66
washer thickness 5

The construction sequence is as follows. Choose an A4 sheet size in landscape style and draw a centre line across the sheet where Y = 100, then add vertical centre lines with X = 50, X = 140 and X = 230. These are the vertical centre lines of the three stages. Draw stage 1, copy it onto the centre line of stage 2, copy stage 2 onto the centre line of stage 3, line in the relevant parts and delete unwanted construction lines.

It is always vital to use the SNAP feature whenever possible to maintain the accuracy of draughting from datum positions, so that MOVE, MIRROR, COPY, and ROTATE can be used with complete accuracy. Select **Polygon** in the **Draw** menu and the command line reads

```
Command: _polygon Number of sides <4>:
```

Type **6** and press <Enter>. The command line now reads `Edge/<Centre of Polygon>:`

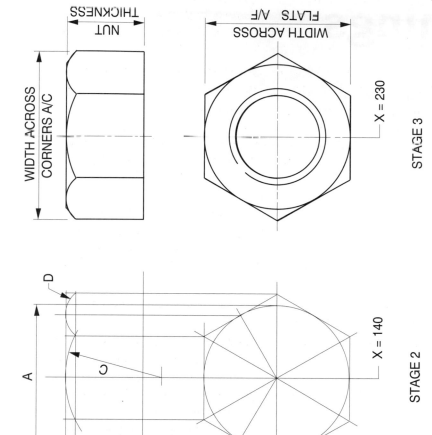

STAGE 3

WIDTH ACROSS FLATS A/F

NUT THICKNESS

WIDTH ACROSS CORNERS A/C

X = 230

NB 1/3

STAGE 2

X = 140

A

B

C

D

STAGE 1

X = 50

Y = 150

Y = 100

Figure 8.1

Click on the required centre where X = 50 and Y = 100 and the command line changes to `Radius of circle:`. Enter **27.5**, which is the distance across flats.

We also need the tangential circle inside the hexagon so select **Circle** in the **Draw** menu, choose **Centre Radius,** click on the centrepoint and insert **27.5**. Draw another datum across the sheet with Y = 150, mark off the nut thickness of 29 and project up the four corners of the hexagon. Clean up the linework and leave as shown.

Commence stage 2 by using the COPY feature to transfer the first stage 90 mm to the right; note the accuracy that the SNAP feature will guarantee.

The corners of nuts are chamfered and the machining process leaves a radius equal to the thread size, which in this case is 36 mm on the front face. Draw a horizontal line across the upper view where this arc intersects with the four projected lines from the hexagon corners. The profile at the top of the nut has four components, namely (referring to Fig. 8.1) the flat top A, the chamfered corners B, the arc on the front face C and arcs projected from the receding faces D.

COPY will transfer the construction to stage 3. Line in the outlines using polylines to give the contrasting line thicknesses required by BS 308. The plan requires the additions of two circles representing the major and minor thread diameters, and these are 36 and 31 mm respectively. It is a BS standard convention to show a female thread with a broken line on the outside circle. Delete unwanted lines to complete stage 3.

SAVE your drawing after checking for accuracy. SAVE your drawing again with a different number and this will now be used in the next example.

Figure 8.2 shows constructions for a nut and bolt assembly and uses parts of the stage 3 finished views. Delete the stage 1 and 2 constructional drawings and other notes. Rotate the two views through 90° in the anticlockwise direction using the centre point of the nut as the axis. Move the two views into the positions shown on stage 4. If you have used the SNAP feature during these changes, the repositioning will be exact.

Nuts, bolt heads and locknuts have different thicknesses, but the profiles which include the constructed radii are identical and these are the features which CAD allows us to reproduce rapidly. Add the construction lines shown for the bolthead, nut, locknut and also two washers. Use the MIRROR command about axis MM to position the nut. COPY the profile at the end on the nut to give the right-hand side of the locknut. Now MIRROR this profile, using axis M2–M2 to give the other side. In the plan view of the assembly only two faces of the hexagon are visible, but construction follows the same procedure after inserting the two arcs. The end of the bolt is either chamfered or radiused, with an arc equal in size to the thread diameter.

Figure 8.3 shows the bolt assembly in first angle projection, so looking from left to right you will see the outside diameter of the washer. Delete all unwanted lines.

Draughting conventions associated with nuts and bolts

Figure 8.4 shows a summary of terms related to nuts and bolts. Table 8.1 gives the dimensions of a selection of thread sizes in common use, and these are covered by ISO 272 and BS 3692. Table 8.2 shows the dimensions of a selection of suitable washers from BS 4320. This Standard gives two thicknesses of washers which may be plain or chamfered at 30°. It is the custom on all drawings to draw only outside views of ordinary nuts, bolts, screws and washers.

The lengths of bolts and screws are always taken from under the head, with the exception of countersunk screws which include the conical head. The plan view of a female thread, for example of a nut, is shown by two concentric circles to represent the major and minor

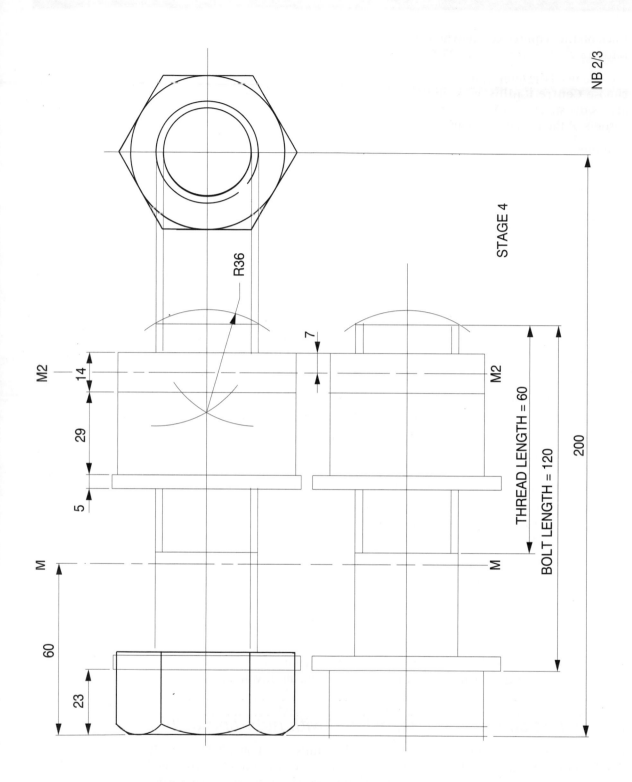

STAGE 4

M2

14

R36

29

5

7

M2

THREAD LENGTH = 60

BOLT LENGTH = 120

200

M

M

60

23

Figure 8.2

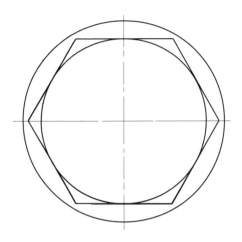

STAGE 5
M36 BOLT ASSEMBLY

NB 3/3

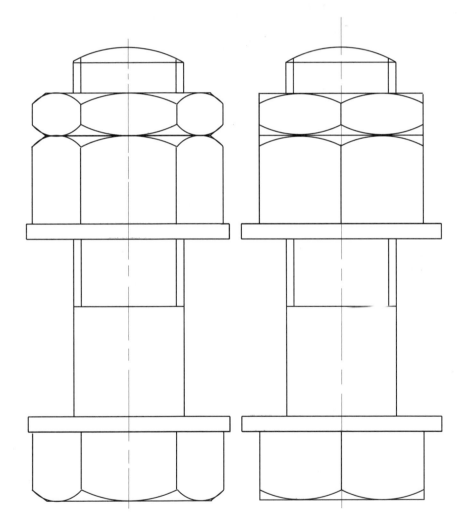

Figure 8.3

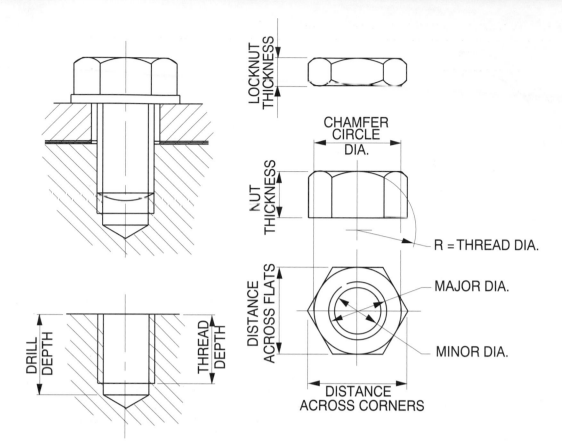

Figure 8.4

Table 8.1 Dimensional details of nuts and bolts; the figures from the appropriate standards have been rounded up to the nearest 0.5 mm for convenience of CAD drawing

D	A	B	A/C	A/F	H	T	t
M1.6	0.35	1.1	3.5	3.0	1.0	1.25	
M2	0.4	1.4	4.5	4.0	1.5	1.5	
M2.5	0.45	1.9	5.5	5.0	1.75	2.0	
M3	0.5	2.3	6.0	5.5	2.0	2.25	
M4	0.7	3.0	8.0	7.0	2.75	3.0	
M5	0.8	3.9	9.0	8.0	3.5	4.0	
M6	1.0	4.7	11.5	10.0	4.0	5.0	
M8	1.25	6.4	15.0	13.0	5.5	6.5	5.0
M10	1.5	8.1	19.5	17.0	7.0	8.0	6.0
M12	1.75	9.7	21.5	19.0	8.0	10.0	7.0
M16	2.0	13.5	27.0	24.0	10.0	13.0	8.0
M20	2.5	16.7	34.0	30.0	13.0	16.0	9.0
M24	3.0	20.0	41.5	36.0	15.0	19.0	10.0
M30	3.5	25.5	52.0	46.0	19.0	24.0	12.0
M36	4.0	31.0	62.5	55.0	23.0	29.0	14.0

Notes

A, thread pitch; D, nominal size thread diameter; B, minor diameter of thread; A/C, width across corners; A/F, width across flats; H, height of bolt head; T, thickness of normal nut; t, thickness of thin nut.

Table 8.2 Dimensional details of thick and thin washers

	A	B	C	D
M1.6	1.7	4.0	0.3	
M2	2.2	5.0	0.3	
M2.5	2.7	6.5	0.5	
M3	3.2	7.0	0.5	
M4	4.3	9.0	0.8	
M5	5.3	10.0	1.0	
M6	6.4	12.5	1.6	0.8
M8	8.4	17	1.6	1.0
M10	10.5	21	2.0	2.5
M12	13.0	24	2.5	1.6
M16	17.0	30	3.0	2.0
M20	21.0	37	3.0	2.0
M24	25.0	44	4.0	2.5
M30	31.0	56	4.0	2.5
M36	37.0	66	5.0	3.0

Notes

A, washer inside diameter; B, washer outside diameter; C, washer thickness Form A; D, washer thickness Form B.

diameters of the thread. The outside circle has a small break in the circumference. The plan view of a male thread, for example of a bolt, is shown by circles representing the major and minor diameters but the inside circle has a small break in the circumference. The thread form is shown by parallel lines representing the major and minor diameters.

In sectional views where threads are drawn in engagement, it is not the custom to cross-hatch between the parallel lines. Empty threads are cross-hatched to the minor diameter and examples of these conditions are given. The profile at the top of a nut is due to the manufacturing process of chamfering the corners. Locknuts are roughly half the thickness of full nuts and are often chamfered on both sides. A bolthead is also chamfered but its thickness is less than that of a nut with the same screw thread. For exact dimensions refer to the appropriate standard tables.

Stud assembly in tapped hole

The following notes relate to the assembly drawing in Fig. 8.5.

- Only the outside views of nuts, washers and studs are drawn in sectional views.
- Where very thin components are used in a sectional view it is customary to block them in rather than hatch them. The 1 mm thick gasket is shown resting on the engine block.
- Note the clearance hole required to assemble the cylinder head above the gasket.
- The stud, shown separately, is screwed into the block for its whole length, so the thread appears to be flush with the top of the tapped hole.
- The screw thread in the block needs to be deeper than the thread on the stud to provide clearance.
- The block is drilled to the minor diameter of the thread and clearance for the thread-forming operation is provided at the bottom of the hole.
- The depth of the hole is measured from the surface to the bottom of the parallel portion. After the drilling operation, the drill also leaves a conical part at the bottom of the hole with an included angle of 120°.
- Hatching does not cover the area where threads are engaged.

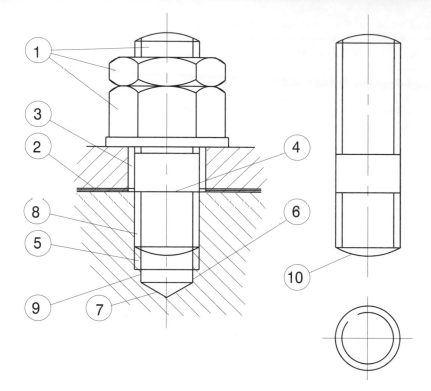

Figure 8.5

- Hatch to the minor diameter where no thread assembly takes place.
- The end of the stud may be finished with a chamfer or radius. The radius is equal in size to the outside diameter of the thread and to the large radius on the face of the nut. Assume that the length of the stud is the usable part which does not include the radiused portion.

Exercise 1: draw a sectional view of a stud assembly in a tapped hole given the following dimensions

- M20 stud: length 65 mm, threaded 35 mm at one end and 17 mm at the other;
- M20 tapped hole: thread depth 24 mm;
- component 13 mm thick;
- 1 mm thick gasket;
- M20 nut and locknut;
- depth of tapping drill 28 mm.

Isometric view of nut

Figures 8.6–8.8 shows the stages in drawing an isometric view of a large nut which could be found on an exploded view of an assembly. Car service manuals use exploded views. For this example use the dimensions for an M30 nut in Table 8.1 and set out the drawing at twice full size. For accuracy it is often more convenient to draw to a large scale and adjust the print size according to requirements.

Stage 1 shows the construction lines. I often find it convenient when transferring distances from one view to another to use circles, as this stage shows. After drawing the first circle the computer remembers its radius, and this is accurate for repetition work; they can easily be deleted later. The dimensions used are as follows:

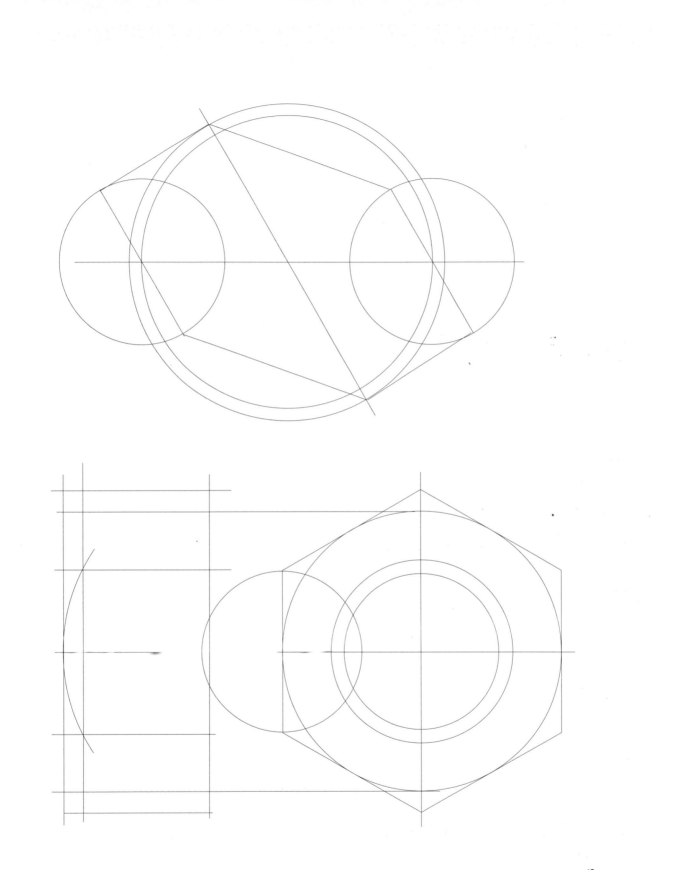

Figure 8.6

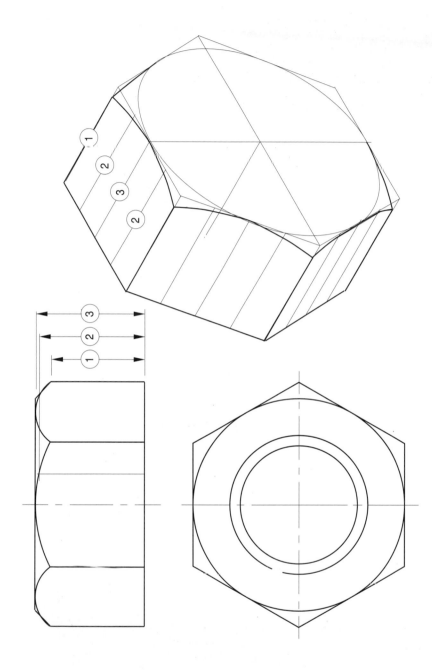

Figure 8.7

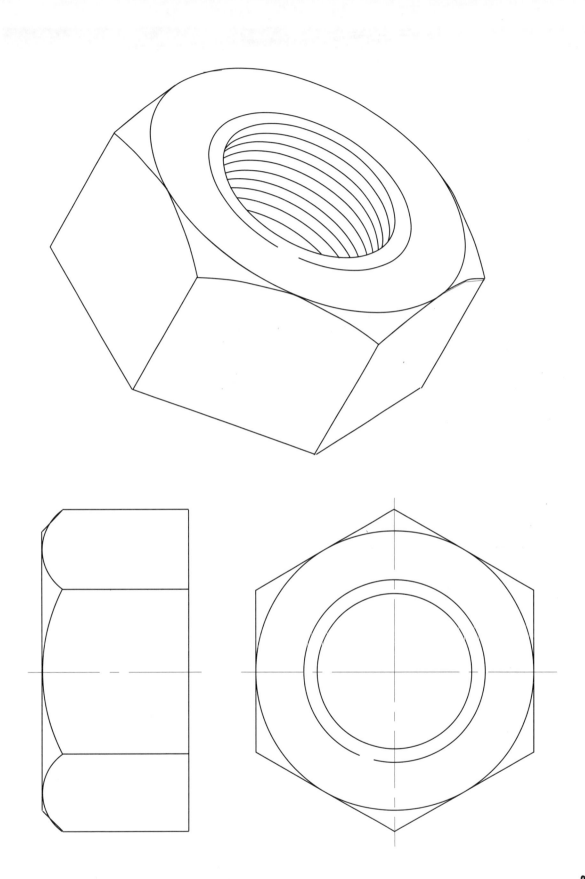

Figure 8.8

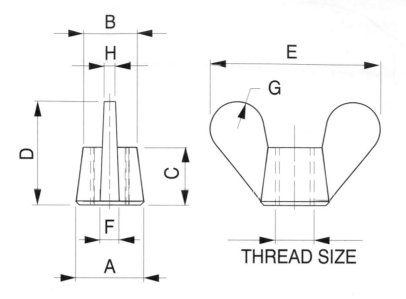

STANDARD DIMENSIONS OF WING NUTS

Figure 8.9

distance across corners 104
distance across flats 92
major diameter of thread 60
minor diameter of thread 51
nut thickness 48
chamfer radius on side of nut 60

Copy the isometric hexagon at an angle of 30° to obtain the rear profile, and line in with polylines. Dimension 1 (Fig. 8.7) determines the length at each corner. The width of the top face is a true dimension but the widths on the two side faces are larger and smaller. However, the ordinates on all faces are true measurements. Position line 2 centrally between the centre lines and their corners and mark off their lengths. The points from ordinates 1, 2 and 3 can be lined in with a polyline then edited into a curve by using the FIT option in the EDIT POLYLINE command.

Draw isometric ellipses for the major and minor thread diameters on the front face. The

Table 8.3 Dimensions of wing nuts in general use (refer to Fig. 8.9)

Thread size	A	B	C	D	E	F	G	H
M3	9	6.5	7	13.5	22	2.5	3.5	1.5
M4 and M5	10	8	9	15	25.5	2.5	4	1.5
M6	13	9.5	11	18	30	2.5	5	1.5
M8	16	12	13	23	38	3	6.5	2.5
M10	17.5	14	14	25.5	44.5	5	7	3
M12	19	16	15	28.5	51	5	8	3
M16	25.5	20.5	19	36.5	63.5	6.5	10	5

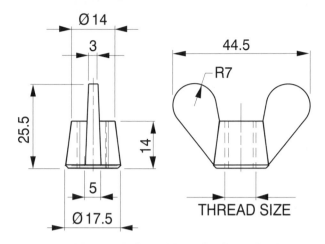

Ø14

3

25.5

14

5

Ø17.5

44.5

R7

THREAD SIZE

M10 WING NUT PROPORTIONS

M6 THREAD
SCALE FACTOR 0.6

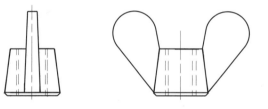

M10 THREAD
FULL SIZE

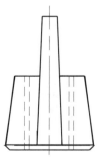

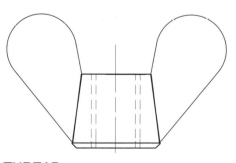

M16 THREAD
SCALE FACTOR 1.6

Figure 8.10

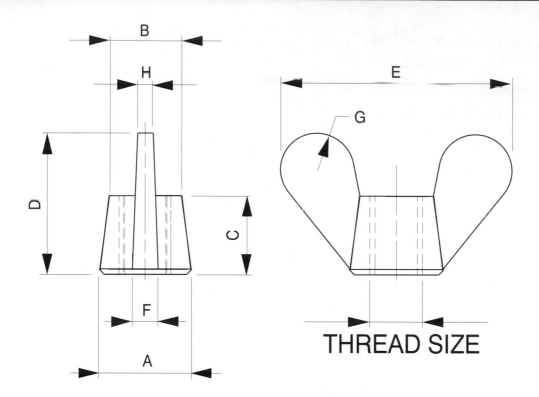

SHOWING EFFECT OF SCALE ENLARGEMENT
ON LINES, ARROWS AND TEXT

Figure 8.11

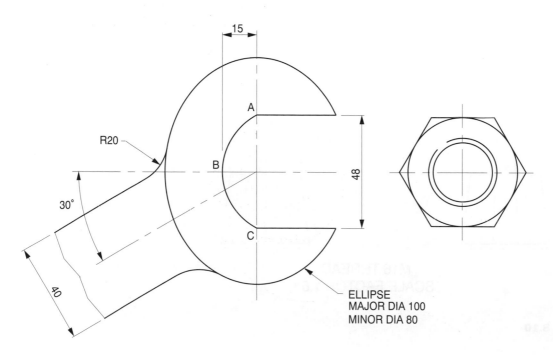

ELLIPSE
MAJOR DIA 100
MINOR DIA 80

Figure 8.12

thread pitch is 3.5 mm, and if the minor thread ellipse is copied at these pitches along the central axis and broken at the boundaries at each end, they will at least represent the roots and crests of the thread helices.

Construction of wing nuts

Table 8.3 and Fig. 8.9 show the dimensions of a range of wing nuts in general use. The SCALE feature can be used to demonstrate the effect of increasing and decreasing the size of drawings. Figure 8.10 shows the dimensions from the table for an M10 wing nut, and a copy of the nut has been made below. The dimensions in the table are graduated approximately in proportion to the thread sizes, so if you require an M16 nut you could take a standard illustration of an M10 nut and increase its size using the scale feature by 1.6. Conversely, an M6 nut will be 0.6 times the size of an M10 nut. These three examples are shown. The result of increasing and reducing sizes means that all aspects of the drawing are changed, for example the line thicknesses, arrows and printing, as shown in Fig. 8.11. The line thickness can always be adjusted by using the POLYLINE feature in the Modify menu, but the arrows and text would need to be replaced.

Nut and spanner exercise

Draw at full size the end of the spanner shown in Fig. 8.12, holding an M30 nut in its jaws.

Construction note

Draw the ellipse as previously described and break the right-hand part for the jaw. Modify the profile to a polyline. ABC is a three point arc from the DRAW menu.

The construction for the R20 radii may present a minor problem. It is not possible to fillet a straight line and an ellipse. The situation can easily be overcome. Draw parallel lines for

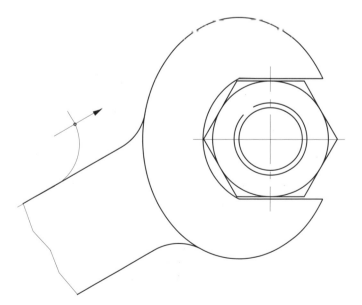

Figure 8.13

the sides of the spanner. Then draw lines perpendicular to them and which are sufficiently long to insert the 20 mm radii.

One line has been drawn on Fig. 8.13. Now turn off ORTHO and SNAP and carefully copy each line to the right at 30° so that it overlaps the profile of the ellipse. I use the COPY feature here because the original line then acts as a guide. If you also use the largest ZOOM frame possible it is quite clear where the profiles of the ellipse and arc overlap. Use the BREAK command to clean up the drawing.

The dimensions for the nut are given in Table 8.1. Slide the nut to the left into position.

Chapter 9

Blocks and technical diagrams

Layouts and diagrams are generally individually prepared special purpose drawings presented in a neat and simple style. If the diagram shows a flow of events then details are arranged so that the sequence, if possible, takes place from left to right or from top to bottom. Notes, dimensions and other references are positioned to avoid ambiguity. Remember that the user may not proficient in that particular branch of technology.

At the end of this chapter you will be able to make a block containing information which can be used in your current drawing or stored for later use elsewhere. Examples of typical diagrams are given for:

- Pipework and heating layouts.
- Electrical schematic line diagrams.
- Electrical wiring diagram for construction purposes.
- Floor layout diagram for a social event.
- Model Hi-Fi Midi sequencer connections.

A diagram shows the relationships between component parts in a scheme, and uses symbols or items of clip art joined together in a logical sequence by lines. Generally, the draughtsman is responsible for the positioning of the various parts and tries to produce a clear layout which is symmetrical and balanced, containing relevant notes and information. As far as possible the linework should be open and not appear as a maze. The CAD COPY, MOVE, and ROTATE features are superb in the respect that, having drawn the main parts of a diagram on the screen you can then reposition them as often as is necessary to improve the spacing.

There are symbols libraries for all branches of technology which are available on disc. Many public libraries also contain copies of British and International Standards, and symbols can be copied or scanned. Make a practice of keeping a file of those types you regularly use. CAD programs use a method of making blocks where complete drawings can be inserted in other drawings and treated as separated units within that drawing. Examples of this follow.

Making blocks

There are two different types of applications for blocks available and an example of each is given here.

1. A block can be written to disc as another drawing file to form part of your library of standard components. This type of block is called a WBLOCK or world block and generally saved in a separate directory. In this case the block can be inserted in any other drawing.
2. Blocks can be made and used for repetition use in the same drawing and the following village hall seating plan is a typical example.

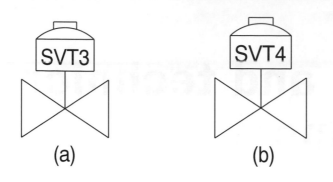

Figure 9.1

Pipework layout application

Assume symbols representing special types of motorised valves are required for insertion on a diagram and one of them is designated SVT3. Fig. 9.1(a) shows a suitable original symbol.

Choose **Make Block** from the **Construct** menu and the Block Definition dialogue box in Fig. 9.2 appears. Add the Block Name and in the Select Point box insert the chosen base reference point for the valve, which in this case can be the centre of the diagonal part of the symbol. This point will be on the pipeline. The intersection of the X co-ordinate of 100 and Y co-ordinate of 150 gives its position on the symbol design sheet.

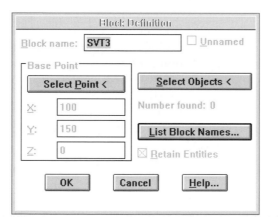

Figure 9.2

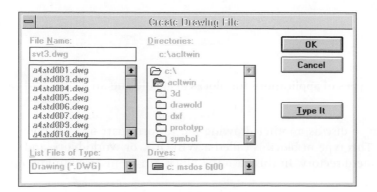

Figure 9.3

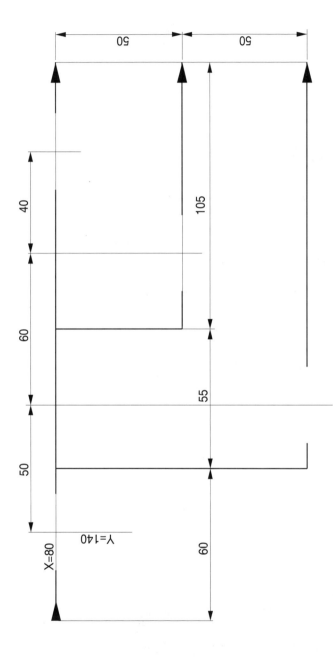

Figure 9.4

Pick the **Select Objects** box. Draw a box around the valve and press <Enter>. Note that the number found in the Block Definition box changes (in this case) to 15. If you wish to check that the block has been successfully made enter **Block** at the command line then **?** and a list of defined blocks will appear.

To define and write the block to disc type **WBLOCK** at the command line or select the **Import/Export** option in the **File** menu. Click on **Block Out**. The Create Drawing File dialogue box appears shown in Fig. 9.3. Enter the symbol name **SVT3** and choose **OK**. The block now has its own drawing file number. The command line asks for the Block name. Enter **SVT3**. Repeat this procedure for the other valve by saving your drawing under a new file name and change the number to SVT4. Make the second block as before.

A typical application follows in Fig. 9.4 where part of a piping layout is shown and it is necessary to position accurately two valves of each type. The co-ordinates for valve SVT3 are X = 80, Y = 140 and X = 130, Y = 40, and those for valve SVT4 are X = 230, Y = 140 and X = 190, Y = 90.

The next operation is to insert the symbol in the correct position on the line diagram. Type **DDINSERT** at the command line and the Insert box in Fig. 9.5 appears. Click on the block box and insert **SVT3**. A choice is offered for you, either (1) to specify the insert parameters on the screen, or (2) to delete the tick mark in the Specify Parameters on screen box, and insert the X value of 80 and the Y value of 140 in the Insertion Point section. Make your choice and click **OK**. Note that you have the option of changing the scale in the X and Y axes, but both must be the same or distorted images will result. The symbol may be rotated by inserting an angle. The EXPLODE button here enables you to break the block into its component parts and make modifications.

In another example we wish to draw a table layout plan for a social function at a village hall to accommodate four tables each with seven seats and 10 tables each with five seats. Part of this exercise is also to prepare all of the drawings on a single A3 sheet.

Set the drawings limits to 420,297 (Fig. 9.7). The plan view of one seat is drawn to a scale of 1:10 and dimensioned from the centre of the table. Copy this drawing, delete the dimensions and use the POLAR ARRAY feature to construct a group with seven seats. The

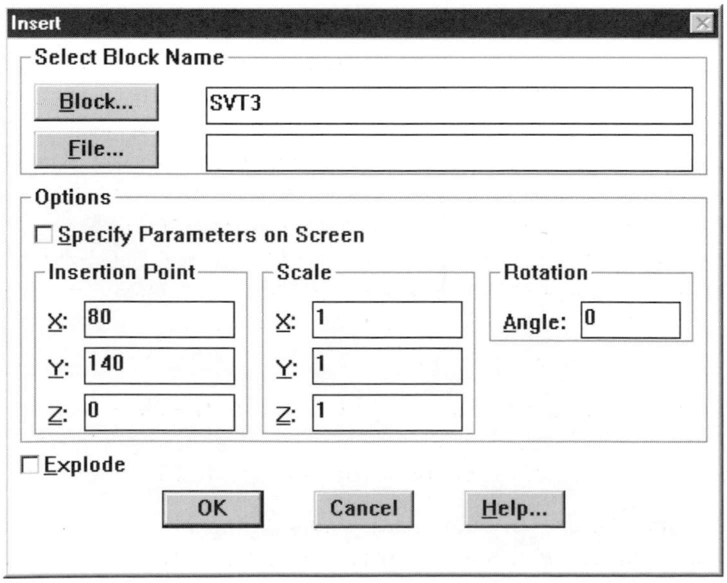

Figure 9.5

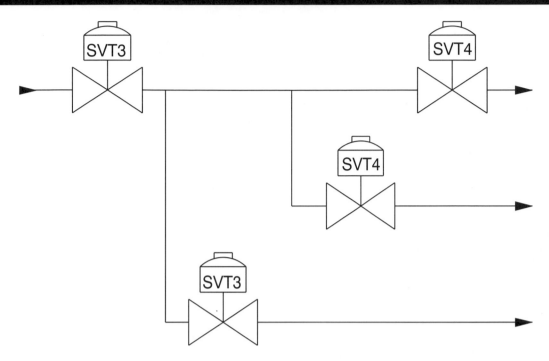

Figure 9.6

area taken by this drawing is large, so use the SCALE feature with a value of 0.5 to give a scale reduction of 1:2.

Repeat with the other five-seat layout. Move the two drawings to the right of the screen; this will leave a space on the left for the hall floor plan. It is necessary to make blocks for the two different groups. For each of them in turn, type **BLOCK** at the command line, choose **Make Block** from the **Construct** menu for the Block Definition dialogue box as before. I gave the name of 7TB for the larger group and the co-ordinates on my drawing were X = 260 and Y = 70. The co-ordinates of the smaller group 5TA were X = 250 and Y = 180.

The previous example for valves related to drawings prepared on separate drawing sheets. If the blocks are prepared on the same drawing as the final layout, then a slightly different command is now given. Type **DDINSERT** at the command line and the Insert dialogue box appears, where instructions are given for direct insertion into the current drawing. Insert the block name **7TB** and the insertion point X = **100** and Y = **70**. The box also permits a change of scale from the scale used to draw the blocks. The final scale of the hall plan is to be 1:100, so enter **0.2** for the values of X and Y scales. Note that from the original table drawing size there have been two scale reductions, of 1:10 and then 1:2, so this final reduction of 1:5 gives a final size of 1:100. These proportions will permit printing on an A4 sheet.

If necessary the block may be rotated at the time of insertion. Use the dimensions to fix a pair of co-ordinates for one of the other groups. The process may be repeated, or the COPY and ROTATE features may be used separately for the rest of the table positions.

Attributes within a block may be edited, and a simple example could be the addition of table numbers. At this stage, however, it is probably far quicker simply to make headings at the side of the drawing with the TEXT command and then use the MOVE command finally to reposition them on the tables.

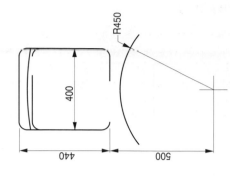

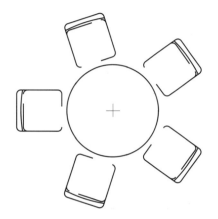

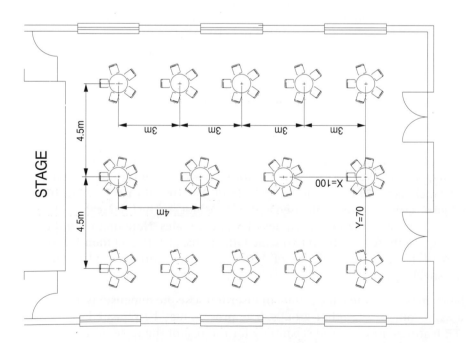

Figure 9.7

Midi sequencer application

It is often necessary to write clear instructions and produce understandable diagrams where products are interconnected, for the benefit of users. Hi-fi and computer controlled equipment are typical examples where several expensive components need to be wired together. All-embracing universal diagrams are sometimes confusing when they are not up-to-date or contain models of equipment slightly different to those you happen to have purchased. A diagram should therefore be accurate, clear, relevant, unambiguous and in good proportion.

The diagrams in Fig. 9.8 emphasise some helpful points. The left-hand line diagram gives the distribution of electronic signals between a computer and musical instruments in a MIDI (musical instrument digital interface) system. The computer's function is as a

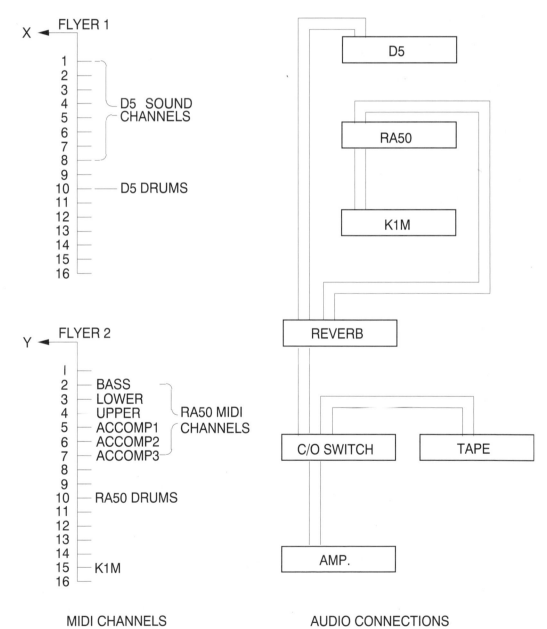

Figure 9.8

sequencer distributing stored musical notes via two output flyers. Flyer 1 is connected from port X to an eight-channel digital keyboard (Roland D5) which also has a built-in drum machine. Flyer 2 is connected from port Y to a music arranger (Roland RA50), which also has its own built-in drum machine, and to a digital synthesiser module (Kawai K1M). This synthesiser can play any one of 128 sounds. All of the instruments play together with backing rhythms.

The diagram is a useful exercise in setting out. Proceed as follows.

1. For Flyer 1, draw the vertical line and on the right side draw 5 mm lines at a vertical pitch of 5 mm. You will find that the COPY feature can be used to copy two, then four, then eight to give the 16 branches.

2. To the left use the TEXT feature in the Draw menu and one above the other draw the numbers 1 to 16. These numbers need to be positioned at the same vertical pitch as the branches, and this can be done using the MOVE command. When the item is selected, click on the centre of the letter and reposition it centrally with the branch line.

3. Reposition the numbers 10 to 16 so that they are in a vertical line with the numbers 1 to 9.

4. The bracket from numbers 1 to 8 can be made from short lines and the FILLET feature. Make one side and MIRROR the other exactly the same.

5. Ensure that the printed notes are positioned centrally and one above the other.

6. Use the ZOOM feature wherever possible.

7. Copy the numbers 1 to 16 and the branch construction below and add the notes applicable to Flyer 2.

The diagram on the right shows the audio connections between the instruments mentioned above together with a reverberation unit, tape recorder, amplifier and a change-over switch for playback purposes. Take particular care in the positioning of the various blocks to ensure that they are symmetrically spaced. Make one block using the RECTANGLE feature and copy the others from it. I always write the text separately, then piece it together and position it centrally within each block. Try to match the style with any other notes appearing on the same, or adjacent, diagrams.

Where parallel lines are required these can be drawn automatically, but I find it easier to use single lines with the SNAP feature for spacing. Their widths should of course be identical. Clean up all corners with the TRIM and EXTEND features.

Heating systems

A schematic diagram of a basic heating system is shown in Fig. 9.9, where a boiler circulates hot water through pipework to radiators and a hot water storage cylinder.

Central heating systems can be configured in many ways, but good thermostatic controls are essential in every case. It is necessary to have flexible controls to turn the system on automatically when required to heat rooms in use but without overheating them. The system must provide adequate quantities of hot water, but not too hot, to prevent scalding.

A house should be well insulated. It is often the case, though, that there will be only one thermostat for the whole house, situated in the hall. If the hallway gets cool when the front door is opened in winter, every downstairs radiator may then emit more heat than is required. Heat rises to already-warm bedrooms, and roof insulation keeps it there. Stiflingly hot bedrooms give rise to excessive fuel bills. Each radiator needs its own local thermostatic radiator control valve (LRV) since bedroom temperatures are invariably required to be lower than those in living rooms. Radiators in uninhabited rooms can also

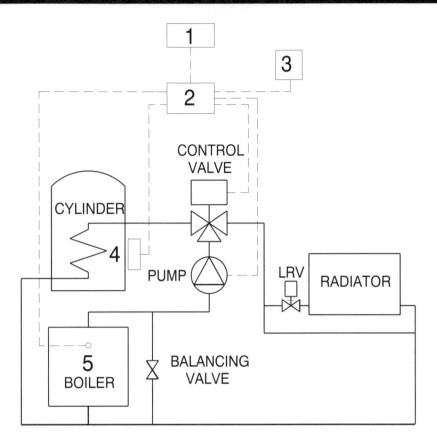

Figure 9.9

be adjusted to a lower level if necessary in order to provide frost protection, and the result will hopefully be that you are only paying for the heat you require.

The system is operated by an adjustable programmer and timer (1) set to suit the lifestyle of the occupants. Each item of electrical equipment is usually wired back to a central terminal board (2). Heating is initiated by a room thermostat (3) responding to falling temperature. Domestic hot water temperatures are controlled at a lower level than those required for heating by a thermostat (4) strapped onto the storage cylinder. When any heat is required, the boiler runs subject to its own control thermostat (5) and built-in high-limit thermostats. The three-way flowshare control valve is part of a flexible program designed to give the occupants a degree of control over their heating system and hot water heating priorities.

The basic scheme here now needs to be expanded and engineered to suit particular applications. The three-way flowshare valve in Fig. 9.10 can be replaced by two separate motorised valves to control the domestic water and heating circuits. Valve X is the water valve, Y the heating valve and Z the balancing valve as before. If either circuit calls for heat, then the boiler will run. Low-water-content boilers require water to be circulating at all times while the burner is alight, and the balancing or bypass valve serves this function. A motorised valve has an internal electrical switch which makes contact when the valve is fully open. If the electrical supply is interrupted, then the valve will close due to the action of a return spring. Note that the balancing valve is set manually by the commissioning engineer when the system is installed.

Copy these schematic diagrams and assume your own proportions for the blocks and symbols. After drawing the first diagram on Fig. 9.9, use the COPY and MOVE facilities to

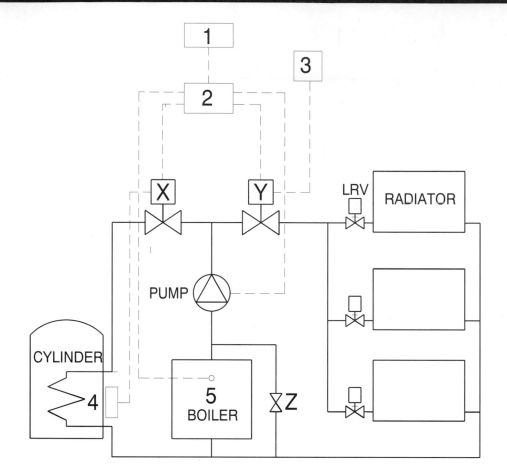

Figure 9.10

reposition the component parts. Drawing modifications form a major part of the draughtsman's weekly work load.

Electrical wiring diagrams for suitable domestic controls fitted on this type of installation are shown later. The true line diagram shows the actual electrical connections, and the electrical construction diagram includes details of how the wiring has beeen routed and constructed on site.

Domestic heating system

The diagram shown on Fig. 9.11 has now been adapted to suit a house heated by a total of 10 radiators. The system uses microbore piping which connects each of the radiators to flow and return distribution manifolds. A manifold is simply a length of tubing with tappings along the side and enables the engineer to run a continuous length of 8 mm small-bore piping to each radiator directly. Since it is flexible, the piping can be run conveniently underneath the floorboards.

The various components are as follows.

1. Boiler fitted on the outside wall in the conservatory. Capacity 65 000 BTU, with balanced flue which draws in combustion air through a concentric stainless steel duct and discharges exhaust air to atmosphere through an external vent.
2. Drain valve at the lowest point in the installation.
3. Pressure relief valve.

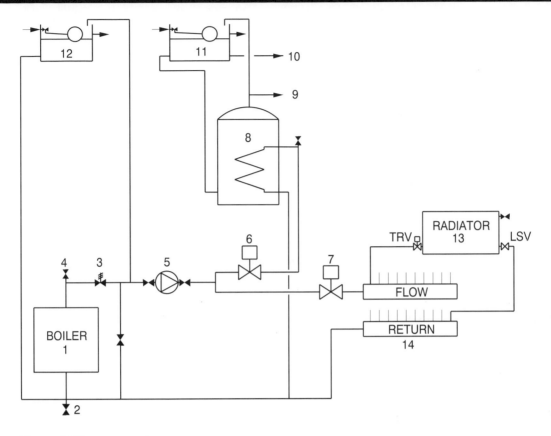

Figure 9.11

4. Air vent to discharge air manually from the system during filling. When filling water systems, air may collect and become trapped at local high points.
5. Pump with isolating valves at the inlet and outlet.
6. Hot water valve which is thermostatically controlled and opened electrically; closed by a return spring.
7. Heating valve with operation similar to the hot water valve.
8. Indirect domestic water cylinder with an air vent at the highest point at the inlet. Often fitted with electric immersion heater in a boss at the top. These are expensive to use and reheat the water relatively slowly. The internal primary coil of pipe is fed with hot water from the central heating boiler via the thermostatically controlled heating valve. The control thermostat is clamped to the shell of the cylinder.
9. Tapping from the top of the cylinder to the hot water connections on the bath, basins and washer. The open expansion pipe terminates above feed tank.
10. Cold water feed from the storage tank to the bath, basins and toilet.
11. Feed tank for the domestic water system. The level is controlled by a ball valve. It is fitted with overflow pipe to the outside of the house.
12. Feed tank to fill the boiler primary circuit at lowest point. An open expansion pipe terminates above the feed tank. An overflow pipe is fitted to the tank. Both feed tanks are connected to the incoming main water supply.
13. Radiator with an air vent at the top. The thermostatic radiator valve provides local temperature control. The lockshield valve controls water flow through radiator.
14. Flow and return manifolds.

Line diagrams

The object of a line diagram is to show the connections in a system to prove its feasibility, but not necessarily wired in the manner in which it will be finally installed. Complete wiring diagrams contain all of the line, neutral and earth connections, and are drawn after taking into account the actual positions of the component parts in the system. They are, in fact, drawn as the equipment is built and possibly include details of wire sizes and colour codings. Manufacturers devise their own methods and presentations to suit individual products and service requirements.

Due to its simplicity in layout, the circuitry in a line diagram is easier to follow. Figure 9.12 shows the line diagram for the central heating system described above. The time control consists of a clock which must run continuously while the circuit is energised. The timer also contains two switches which may be operated independently. One switch controls the central heating circuit, the other the domestic hot water circuit. The controls can be switched on and off to suit family requirements and the time of year. It is customary to draw the line diagram with switches in the normally off position where possible.

At a specified time, the clock makes the contact to terminal 3 on the timer and, if the cylinder thermostat is calling for heat, the brown terminal on the domestic water valve will be energised. Both valves are motorised and when energised will open. When the line is switched off, they close due to the action of a spring. When the cylinder thermostat is satisfied, the circuit is broken.

In many homes the domestic water is automatically switched to the on position morning and evening for a couple of hours, but an override switch also provides hot water control

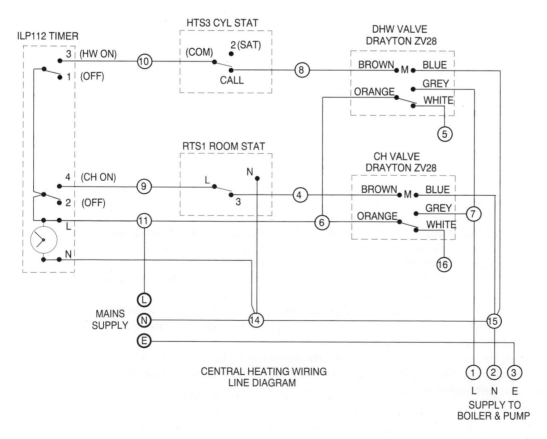

CENTRAL HEATING WIRING
LINE DIAGRAM

Figure 9.12

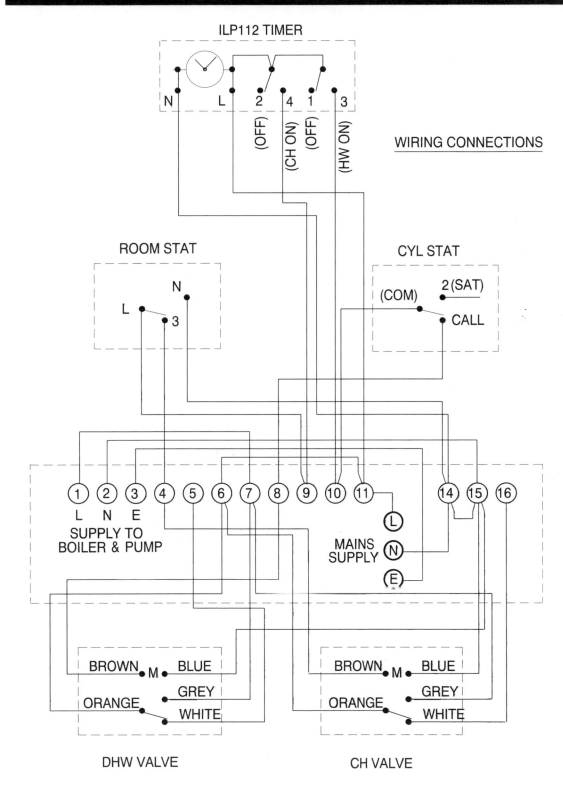

Figure 9.13

at all times. On the central heating circuit, the line from the timer contact 4 will energise the valve and allow water to flow to the radiators, provided the room thermostat is calling for heat. Note that the grey terminals on both valves are connected to terminal 7 because, if heat is required for either system, the boiler must be switched on. Note also that the terminals marked L and N on the room thermostat cannot be shorted out when heat is not required. The thermostat is electronic and is operated by relay. Without the neutral connection the switch will remain open, and when the ambient temperature falls and heat is required, no click would be heard.

The layout of a line diagram should hopefully enable the reader to follow the sequence of events like a book.

Construction wiring

The wiring connections for the installation are shown on Fig. 9.13 Since the components are positioned far from each other, a central terminal is now provided so that wires for each component can be run together from an accessible point where electrical operational checks may be made. The terminal numbers have also been added to the line diagram so that tracing the wires is reasonably easy.

You will see on Fig. 9.13 that the wires from the white contacts on the valves are connected to terminals 5 and 16. Signal lamps connected to these terminals and neutral could indicate whether the valves were open or closed.

In large industrial installations, instrument panels with a wide variety of dials, gauges, indicator lights and controls are individually designed and constructed.

Copy both of these diagrams so that each of them can be printed on A4 sheets of paper. You will find that it is a valuable exercise in planning and layout work.

Chapter 10

Technical drawings for industry

Examples are given here of typical technical drawings and may be copied to demonstrate competence as a CAD draughtsman. The drawings are chosen to provide experience of applications of engineering standards, conventions, principles and practice and include:

- Drawing sheet layouts with title blocks, parts lists and borders.
- Webs and fillets applied to castings.
- First and third angle alternative solutions.
- Assembly drawing from given details.
- Transferring information from one drawing to another.
- Sectional views

This chapter deals with complete component and assembly drawings to current industrial standards. The examples are graded to provide the CAD draughtsman with useful experience and skill in manipulating the main features of AutoCAD 2-D software which must be learned in order to pursue a career in a typical industrial drawing office.

I accept that progress is at first very slow but with experience you will soon speed up. There are often alternative methods of procedure, but we all gradually develop our own individual techniques to present an acceptable result. Fortunately CAD is a tool which gives you many opportunities to try and try again. In order to become more proficient, may I suggest that you time yourself as you work your way through the following drawing examples and note your progress. I would comment that many of these examples could be drawn by hand in about 2 hours. At first, CAD takes longer, so don't despair as speed and efficiency gradually increases with software familiarity and application.

Standard drawing sheets

Established company practice is normally to send business letters to customers on headed sheets of paper. Each company has its own standards regarding layout and content, but it is fair to comment that most headed papers have a lot in common with each other. Drawing sheets are also designed individually but there are certain recommendations from British Standard 308 regarding drawing layouts which are applicable to CAD studies. Drawing sizes are given in Table 10.1.

The exercises in this book are all designed to fit A4 sheets of paper so that they can be reproduced easily and economically on standard black ink plotters or printers. The minimum border width for A4 sheets is 10 mm. Drawings may also be photocopied or microfilmed, and centring marks across the frame are considered helpful. An orientation mark is also provided in the form of an open triangle positioned in the centre of the bottom edge facing the user. A scale bar is positioned symmetrically about the orientation

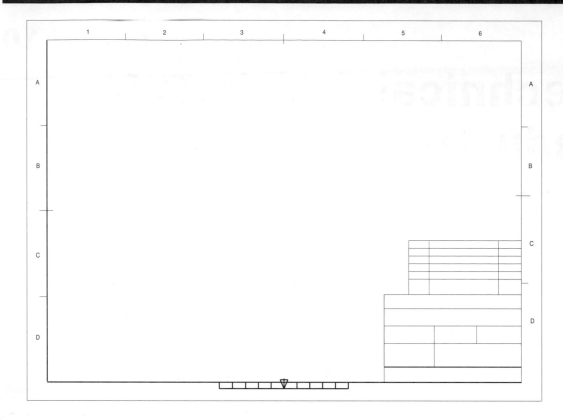

Figure 10.1

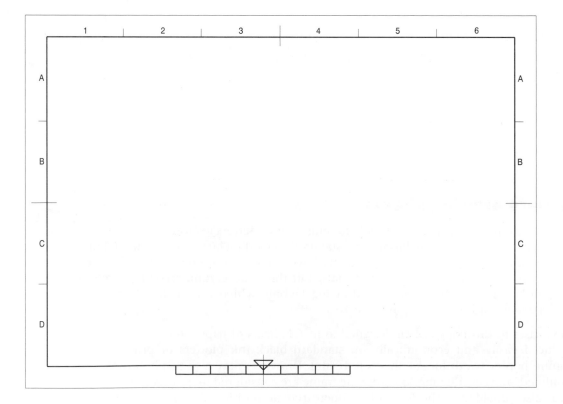

Figure 10.2

Table 10.1 Drawing sizes

Designation	Size (mm)	Minimum border width from drawing frame to edge of sheet (mm)
A0	841 × 1189	20
A1	594 × 841	20
A2	420 × 594	10
A3	297 × 420	10
A4	210 × 297	10

mark, with a minimum length of 100 mm and width of 5 mm, divided into 10 mm intervals.

A grid reference system is desirable for all sheet sizes, so that features and components on the drawing can be located easily for reference during, for example, assembly, servicing and modification applications. A title block is generally positioned at the bottom right-hand corner of the drawing and includes information for identification and other relevant details which may vary according to company practice.

In the case of college work, the examiner will probably require the block to contain the college name, the student's name, the drawing title, number, scale, projection angle and the units of measurement.

Where an assembly drawing is involved, the student may also be required to include a parts list. The numbered components are generally inserted in the box in order of importance. In a separate column the quantity of each component required for that particular assembly is stated. Parts lists are generally positioned at the right-hand side of the drawing immediately above the title block.

It is also customary for the student to add numbered identification balloons to identify each component. A leader line directed from the circle centre, but only commencing at the circumference, terminates in a dot on the component.

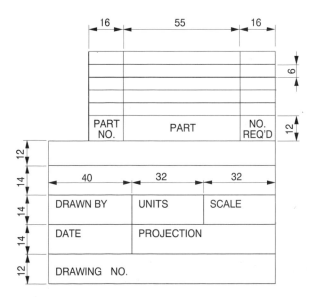

Figure 10.3

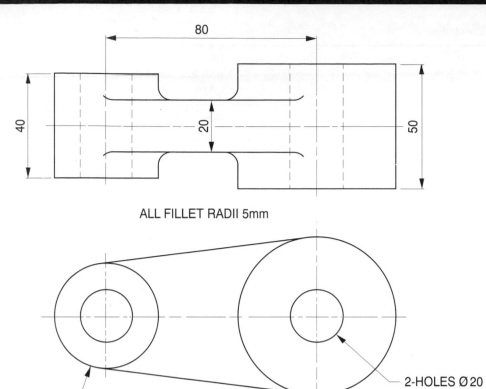

ALL FILLET RADII 5mm

Figure 10.4

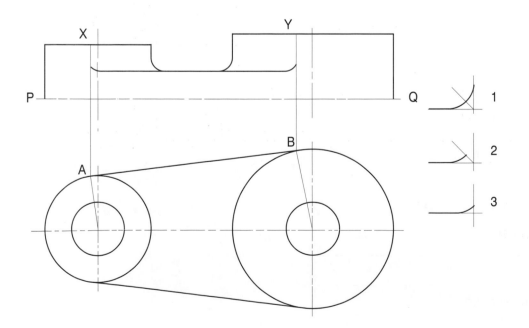

Figure 10.5

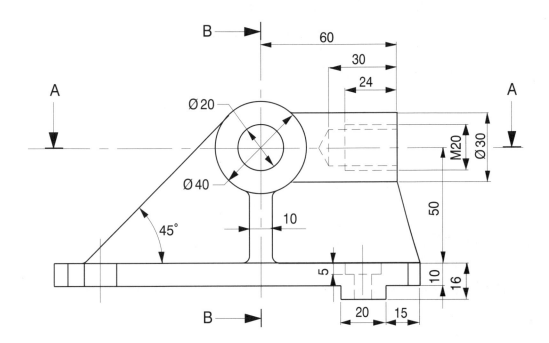

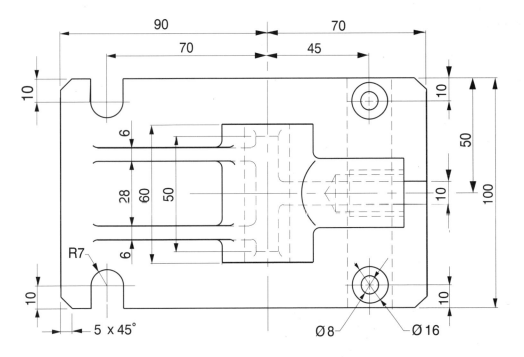

ALL FILLET RADII 3mm

Figure 10.6

In the case of industrial applications the title box will be far more detailed, and typical examples are included in BS 308 to illustrate the scope of the content. Drawings are all numbered and dated, but dates not only state the original issue date but also the dates when various modification took place. A drawing therefore may be issue A, B, C, D etc.

Drawing procedure

Prints cost money and take time, so it is not intended that every drawing should initially be undertaken on a sheet with a professionally finished border and block. A better system is to keep the outline frame and drawing separate until all preliminary work and editing is complete and then join them up to complete the project. The scale features in CAD allow you to take advantage of the space available so that the biggest possible image can be fitted within the frame.

An A4 sheet is satisfactory for some good geometrical constructions, small assemblies and component details. Draw the A4 sheet shown in Fig. 10.1.

An A3 sheet is sufficiently large for every assembly drawing in this book and can easily be reduced in scale and printed on an A4 sheet. Set out an A3 sheet complete with a title block and a parts list as shown in Fig. 10.2. The dimensional details of the block are detailed separately on Fig. 10.3.

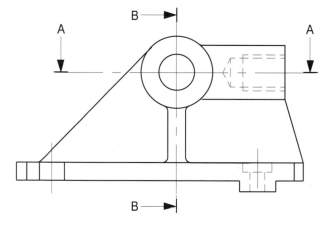

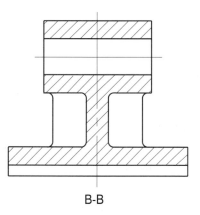

FIRST ANGLE
SOLUTION

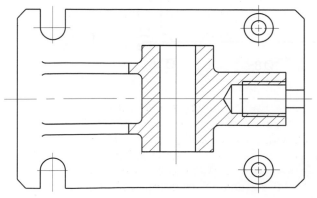

A-A

Figure 10.7

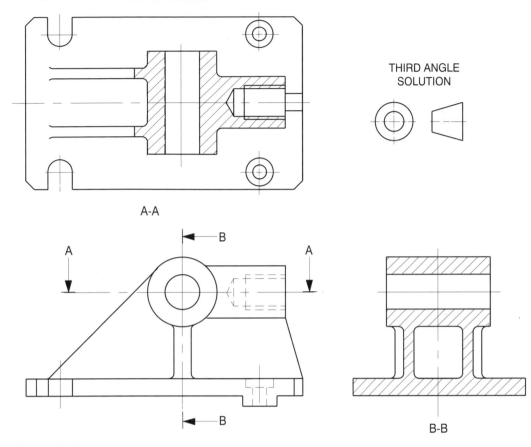

Figure 10.8

The details within the block may be to your own design, but Fig. 10.3 shows typical contents. Some manufacturers position parts lists at the top right-hand corner of the sheet with other relevant notes underneath. Another practice is to quote the drawing number again at the top right-hand corner so that it is visible in a filing system. There are no hard and fast rules here.

It is also a good idea to keep the details of the block on a separate file, to be used in whole or in part as required. The block can be imported into other drawings using the DXF feature, described below and then edited.

Prepare these standard sheets in landscape style as this is the orientation generally used. Standard sheets in portrait style are of course also used when the shape of the drawing demands it.

Drawing webs and fillets

Figure 10.4 shows the dimensions of a mechanical lever which is manufactured by the casting process, and illustrates constructions for drawing webs and fillets. The sand casting process leaves fillet radii between adjacent surfaces, and the draughts-man may add a general note on the drawing when many radii occur: 'All fillet radii 5 mm' is a typical example. This lever consists of two cylindrical bosses with a central web. After casting, the component is machined to ensure that top and bottom surfaces of the bosses are flat and parallel to each other. The two perpendicular holes are then drilled.

Commence by drawing concentric circles in the plan and project the sides up to the front

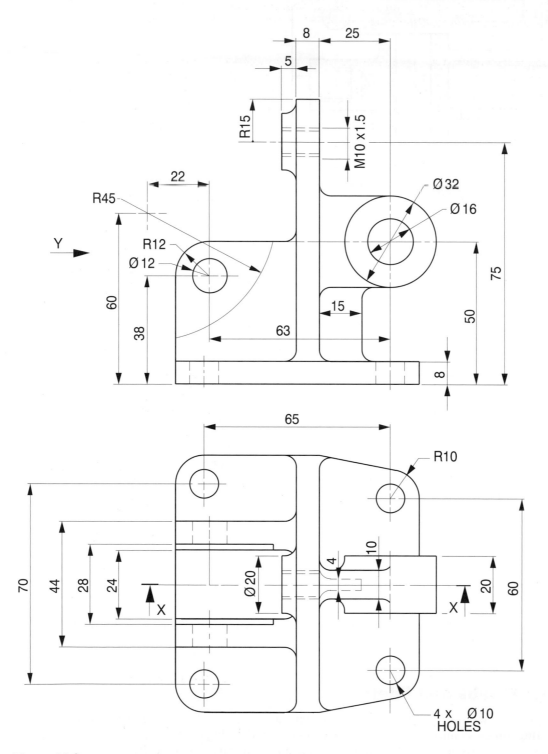

Figure 10.9

view. Add horizontal construction lines to establish the thicknesses of each of the cylindrical parts and the web. Remember to use the ORTHO and SNAP modes whenever possible.

The next stage is to draw tangents to the outside diameters of the bosses and determine the tangency points. AutoCAD conveniently provides OBJECT SNAP modes in the toolbox. Choose the TANGENT button in the toolbox. Follow the instruction at the command line to snap on one of the circles, then the other, and the tangent will be automatically drawn. We now need to establish the exact points of tangency (A and B) at each end of the tangent line (Fig. 10.5).

To position point A, choose the **Line** button in the toolbox. Then choose the **Centre** button in the toolbox and the cursor will change in appearance to a target box. Choose the **Circle** button, click on the circle, and the starting point for the line snaps to the circle centre.

Choose the **Perpendicular** button and click on the tangent on your drawing; a line will be drawn to the exact position. Now we need to project a vertical line up to the front view, and this will be perpendicular to the top face of the boss. Choose the **Perpendicular** button again and click on the face of the boss to give the point X. Repeat the procedure for points B and Y.

You will note on the front view that two 90° fillets are drawn at the centre of the web, but

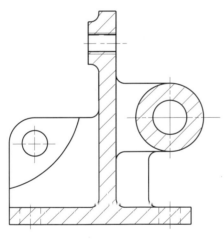

X - X

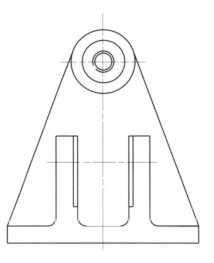

FIRST ANGLE SOLUTION

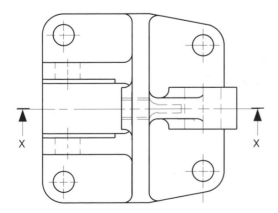

Figure 10.10

at the tangency points, where the projection of the fillet radii terminates, it is the custom to draw only a 45° fillet. Use the FILLET command to insert the four fillets. At each tangency point draw short 45° lines and with the BREAK command erase the unwanted portion. Extend the lines with the STRETCH command.

These constructions are shown in stages 1, 2 and 3. The view can be completed by using the MIRROR command about the axis PQ. Add all dimensions to your drawing.

The following examples illustrate typical castings, manufactured in a foundry, where molten metal is poured into prepared moulds. After cooling, the casting is taken to a machine shop and the surfaces are machined to their final dimensions. The accuracy and the finish of machined surfaces are determined by the designer, and this information is added to complete the working drawing. Dimensioned drawings to describe shape and form consist of outside views and sections through various selected features in order to define the product completely. It is often the practice to draw separate drawings for the foundry where only data applicable to that part of manufacturing is included; a second drawing is then produced showing only details of the work to be performed in the machine shop.

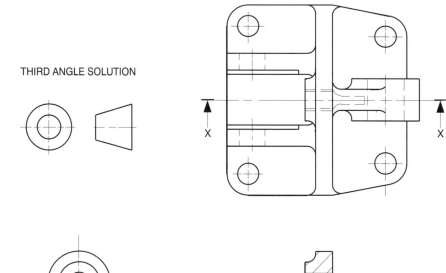

THIRD ANGLE SOLUTION

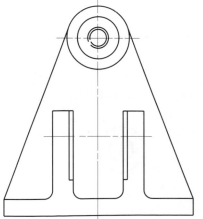

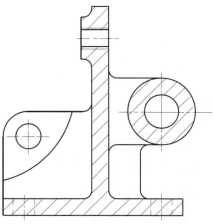

X - X

Figure 10.11

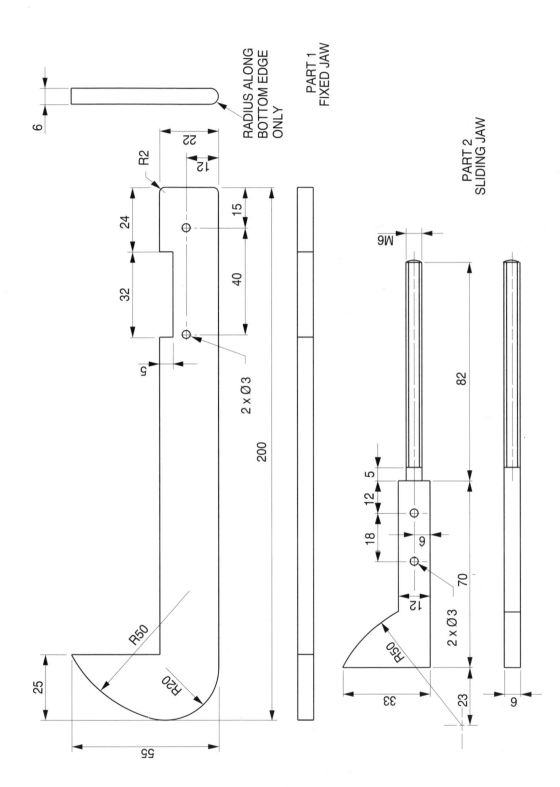

RADIUS ALONG
BOTTOM EDGE
ONLY

PART 1
FIXED JAW

PART 2
SLIDING JAW

Figure 10.12

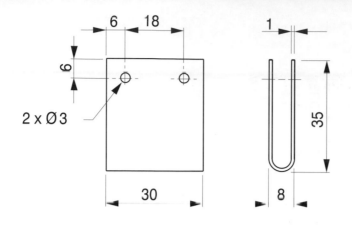

PART 4
SLIDING BRACKET

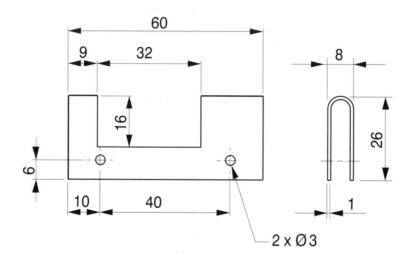

PART 5
FIXED BRACKET

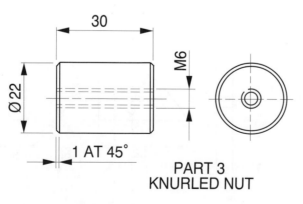

PART 3
KNURLED NUT

Figure 10.13

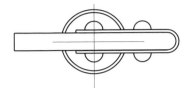

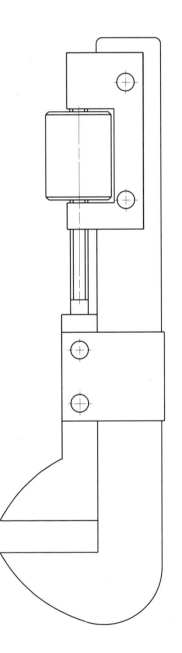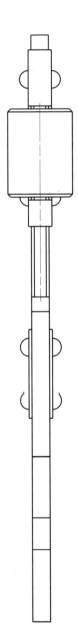

Figure 10.14

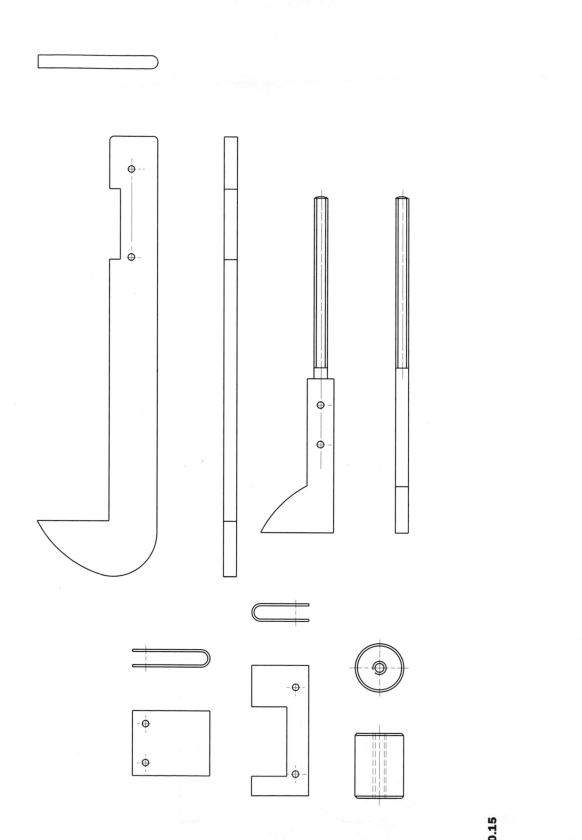

Figure 10.15

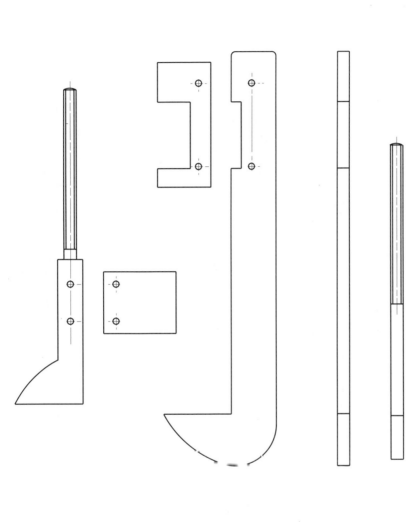

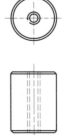

Figure 10.16

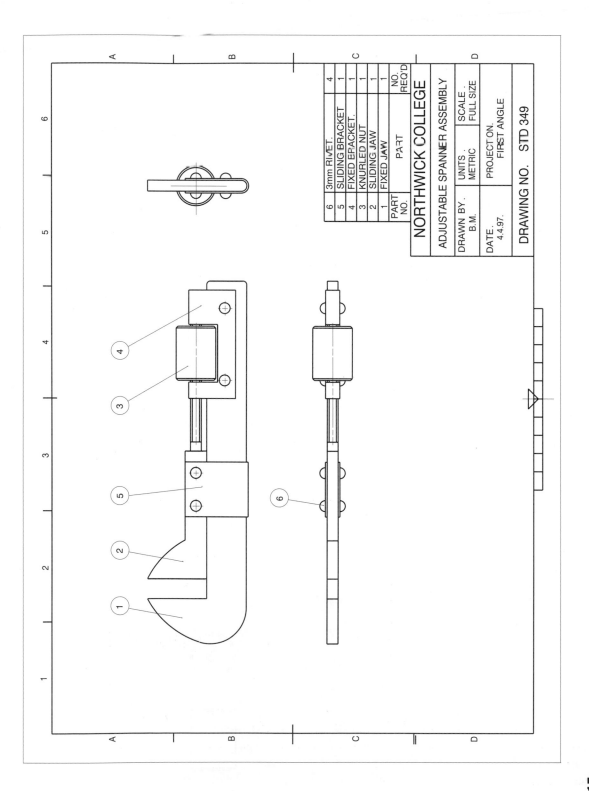

6	3mm RIVET.	4
5	SLIDING BRACKET	1
4	FIXED BRACKET.	1
3	KNURLED NUT	1
2	SLIDING JAW	1
1	FIXED JAW	1
PART NO.	PART	NO. REQ'D

NORTHWICK COLLEGE

ADJUSTABLE SPANNER ASSEMBLY

DRAWN BY. B.M.	UNITS . METRIC	SCALE . FULL SIZE
DATE. 4.4.97.	PROJECT ON. FIRST ANGLE	

DRAWING NO.   STD 349

Figure 10.17

Figure 10.6 shows the front and plan view of a cast anchor bracket in first angle projection. We are required to draw the given front view and project in first or third angle projection the following views:

- a plan view taken as a section on the cutting plane AA;
- a sectional end view on the cutting plane BB.

This example is a typical casting where a central boss is supported by four webs. Other details include two counterbored holes in the base, two slots and four chamfered corners at 45°. At the right-hand side of the boss there is a tapped hole. The example is also given to show how these details are dimensioned and section planes are indicated.

Copy this drawing and accurately position all fillet radii. Save your drawing under a different drawing number and then convert it into the given solution including sections A–A and B–B. Note that the cutting plane for section B–B incorporates the BS 308 convention relating to thin webs, which are not cross-hatched. In Fig. 10.7 the solution is given in first angle projection. Fig. 10.8 shows a third angle projection solution, assuming that the same vertical section plane is used but that the arrows point in the opposite direction to show an alternative end view. There are three points to remember:

- the section plane indicates the position where the component is cut;
- the arrows point to the part of the component to be drawn;
- the projection angle dictates where the drawing will be positioned on the solution.

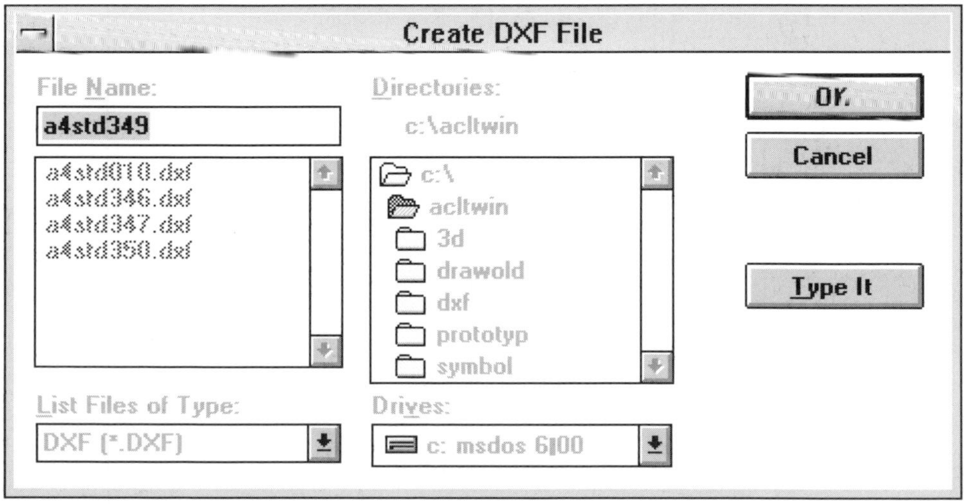

Figure 10.18

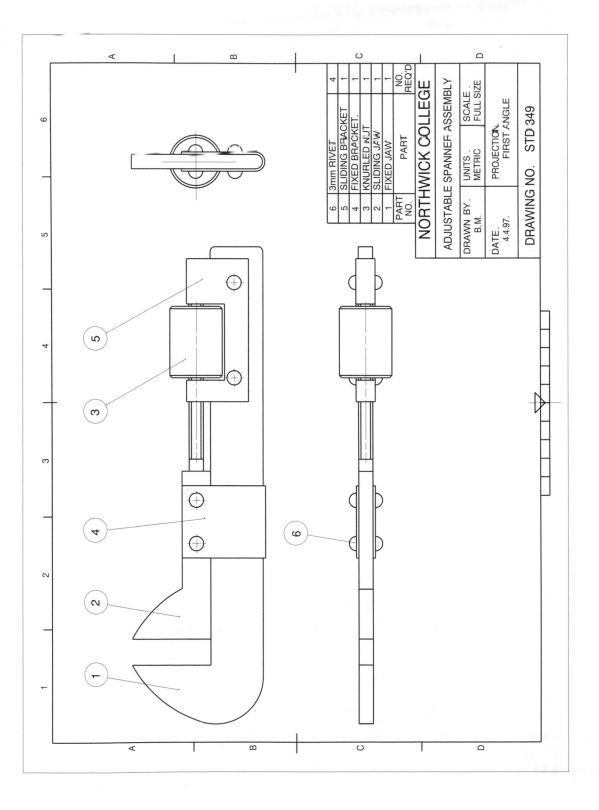

6	3mm RIVET	4
5	SLIDING BRACKET	1
4	FIXED BRACKET.	1
3	KNURLED NUT	1
2	SLIDING JAW	1
1	FIXED JAW	1
PART NO.	PART	NO. REQ'D

NORTHWICK COLLEGE

ADJUSTABLE SPANNER ASSEMBLY

| DRAWN BY. B.M. | UNITS. METRIC | SCALE. FULL SIZE |
| DATE. 4.4.97. | PROJECTION. FIRST ANGLE | |

DRAWING NO. STD 349

Figure 10.19

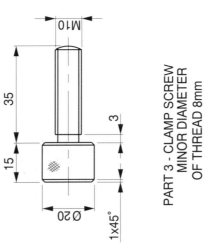

PART 3 - CLAMP SCREW
MINOR DIAMETER
OF THREAD 8mm

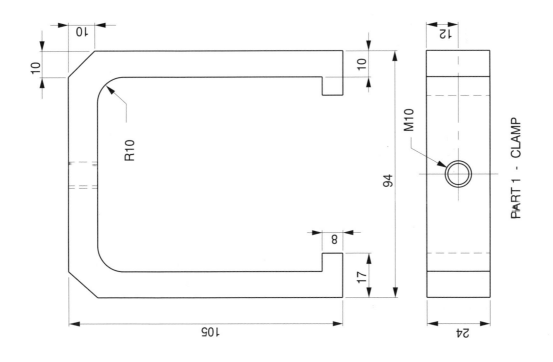

PART 1 - CLAMP

Figure 10.20

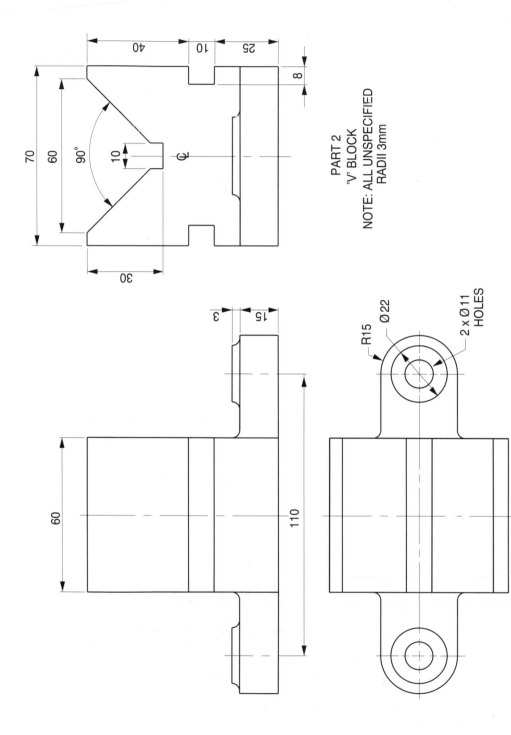

PART 2
'V' BLOCK
NOTE: ALL UNSPECIFIED
RADII 3mm

Figure 10.20 (continued)

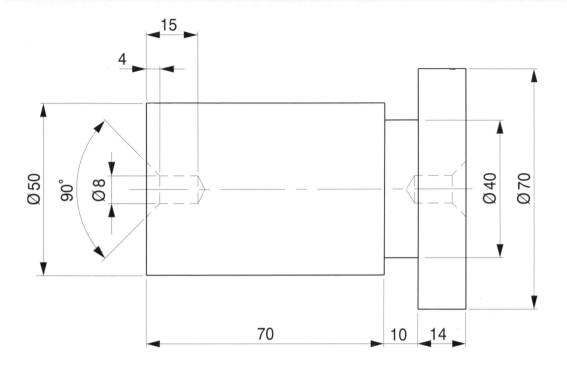

PART 4
COMPONENT

Figure 10.20 (continued)

Cast iron bracket

Figure 10.9 shows the details of a cast iron bracket in first angle projection and we are required to draw the following views:

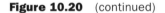

- the given plan view;
- a sectional elevation on plane XX;
- an end elevation looking in the direction of arrow Y.

Solution notes
The section plane cuts through the base and vertical web at right angles, so these details are cross-hatched. It also slices through the two thin webs which support the 32 mm diameter boss on the right-hand side. The convention is not to cross-hatch thin webs and their outlines are defined as shown. Note that the pitch between the 45° lines on the cross-hatching is the same for all areas as we are only drawing one component. The cross-hatching also covers the M10 thread to the minor diameter.

Figure 10.10 shows the solution in first angle, while Figure 10.11 shows the solution in third angle.

Complete assemblies

Examination questions in engineering drawing invariably involve the draughtsman in assembling components together and drawing two or three views in projection. One

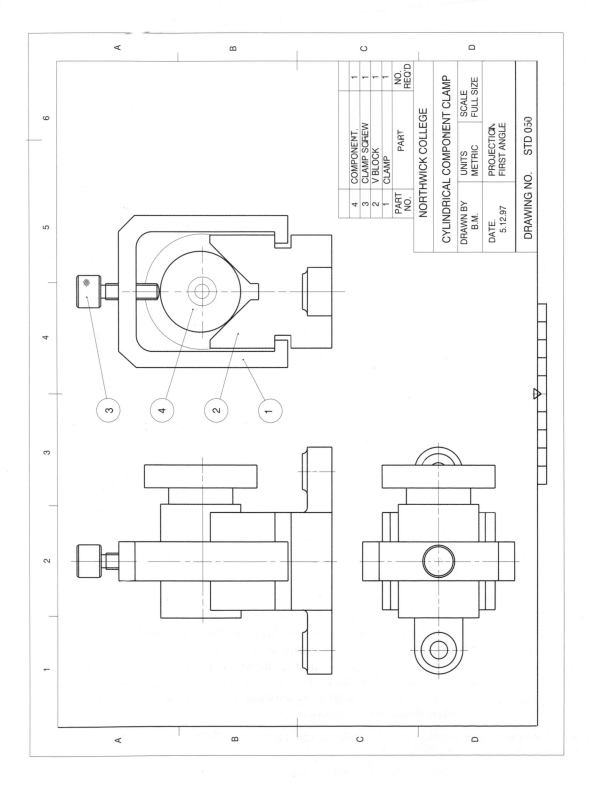

PART NO.	COMPONENT.	NO. REQ'D
4	CLAMP SCREW	1
3	V BLOCK	1
2	CLAMP	1
1	PART	NO. REQ'D

NORTHWICK COLLEGE

CYLINDRICAL COMPONENT CLAMP

DRAWN BY B.M.	UNITS METRIC	SCALE FULL SIZE
DATE. 5.12.97	PROJECTION FIRST ANGLE	

DRAWING NO.   STD 050

Figure 10.21

obviously competes against the clock, so guidance is often provided to assist in the comprehension aspect. The following exercises are complete with solutions to assist in understanding the linework, so that the draughtsman can gain engineering experience and expertise.

Adjustable spanner

The component parts of an adjustable spanner are shown in Figs. 10.12 and 10.13. Copy the given component details and then draw the following views:

- a front view of the assembled parts;
- a plan view;
- an end view.

The components are assembled using four 3mm diameter round-head rivets and the heads of the rivets are shown in the front view as circles of 5mm diameter.

Solution notes

Copy the component drawings and dimension them as shown in Figs. 10.12 and 10.13. Save both drawings on file. Study the assembly solution in Fig.10.14. It can be constructed partly by manipulating the outlines already drawn like a jigsaw puzzle. Note that three-quarters of the assembly is present in Fig. 10.12, so save this drawing using a new file number to give the right-hand side of Fig. 10.15 and delete the dimensions. Increase the drawing limits to suit A3 size, which doubles the size of the drawing sheet. Take a second copy of Fig. 10.13, delete the dimensions and insert it onto Fig. 10.15 giving the result shown. Use the MOVEcommand in the Modify menu to reposition the components as shown in Fig. 10.16.

In CAD there is no point in setting out work twice. It is a relatively simple matter to edit parts 3, 4 and 5 in their correct positions and delete surplus linework. Use ORTHO, SNAP and ZOOM whenever possible. Please refer to the following notes regarding DXF files to arrange the manoeuvre of placing one file into another. The result will be accurate provided the original drawings have been prepared using ORTHO and SNAP for datum purposes. This solution is shown in first angle projection.

Using DXF files

The drawing interchange format (DXF) is a valuable feature of CAD since it allows you to move drawings from one file into another. Furthermore, the files do not have to be in the same drawing software, but can be exported into other Windows applications provided they are compatible; however, this needs to be checked and the importing procedure confirmed at the receiving end.

Assume that the assembly drawing shown in Fig. 10.17 is to be inserted into a frame. Step 1 requires a separate DXF file to be made. Select **Import/Export** from the full **File** menu and a second menu will appear. Click on **DXF Out**. The dialogue box in Fig. 10.18 appears and the file name STD349 (my own file number) is inserted. AutoCAD LT appends `.dxf` to the file name. Click OK and the command line reads

`Enter decimal places of accuracy (0 to 16)/Entities <6>`

We require the entire drawing to be transferred: click **OK**, and the DXF file is made.

Open a receiving file containing a copy of the frame outline to Fig. 10.1 and saved under a new file number. Again select the **Import/Export** option from the full **File**

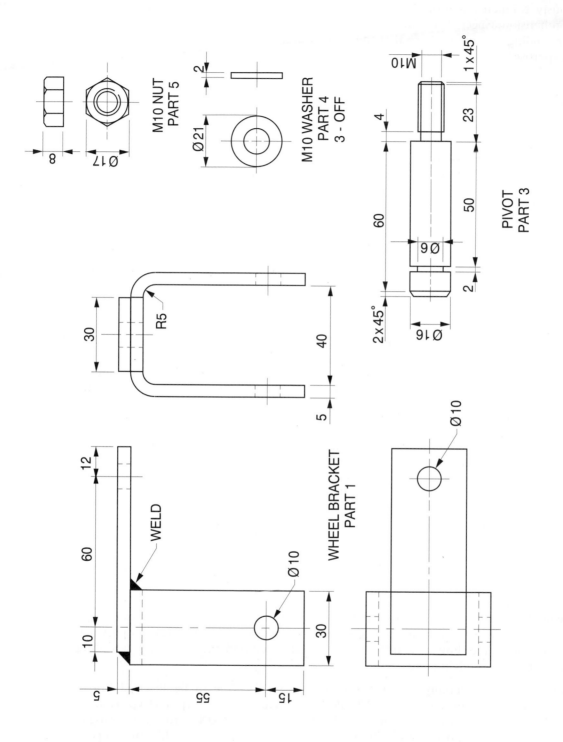

M10 NUT
PART 5

M10 WASHER
PART 4
3 - OFF

PIVOT
PART 3

WHEEL BRACKET
PART 1

WELD

Figure 10.22

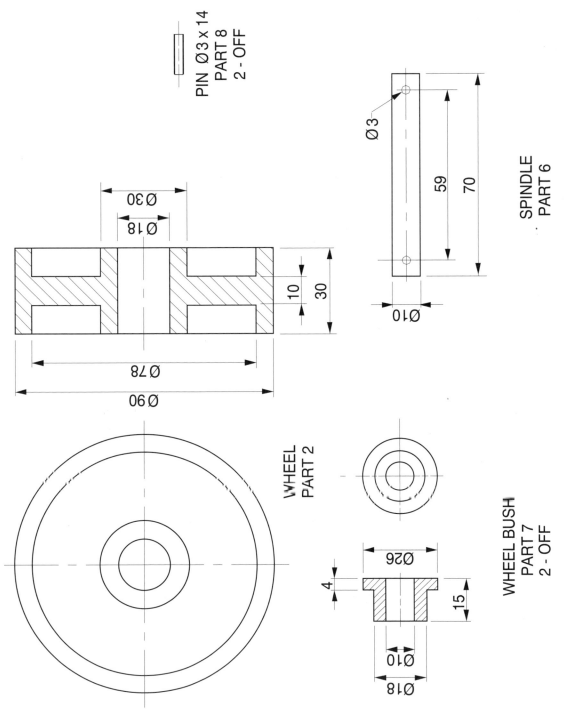

PIN Ø3 x 14
PART 8
2 - OFF

SPINDLE
PART 6

Ø3

59

70

Ø10

Ø30

Ø18

10

30

Ø78

Ø90

WHEEL
PART 2

WHEEL BUSH
PART 7
2 - OFF

Ø26

4

15

Ø10

Ø18

Figure 10.22 (continued)

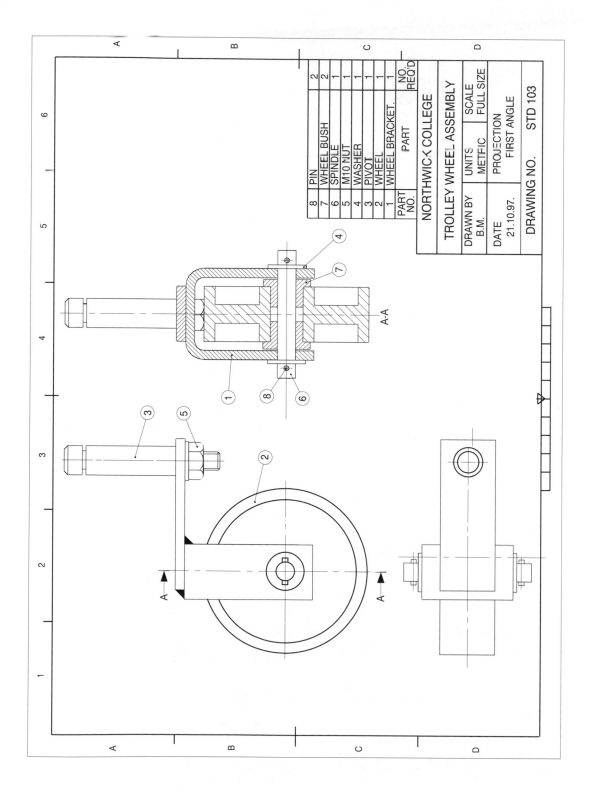

PART NO.	PART	NO. REQ'D
8	PIN	2
7	WHEEL BUSH	2
6	SPINDLE	1
5	M10 NUT	1
4	WASHER	1
3	PIVOT	1
2	WHEEL	1
1	WHEEL BRACKET.	1

NORTHWICK COLLEGE

TROLLEY WHEEL ASSEMBLY

DRAWN BY B.M.	UNITS METRIC	SCALE FULL SIZE
DATE 21.10.97.	PROJECTION FIRST ANGLE	
DRAWING NO.		STD 103

A-A

Figure 10.23

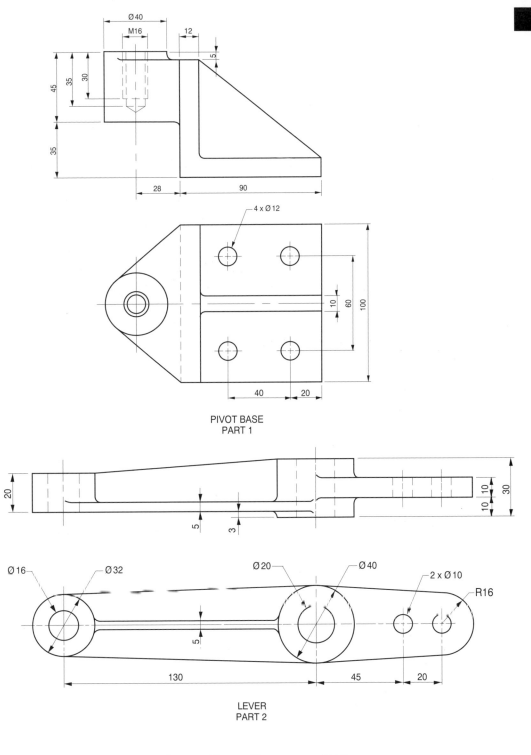

PIVOT BASE
PART 1

LEVER
PART 2

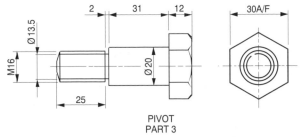

PIVOT
PART 3

Figure 10.24

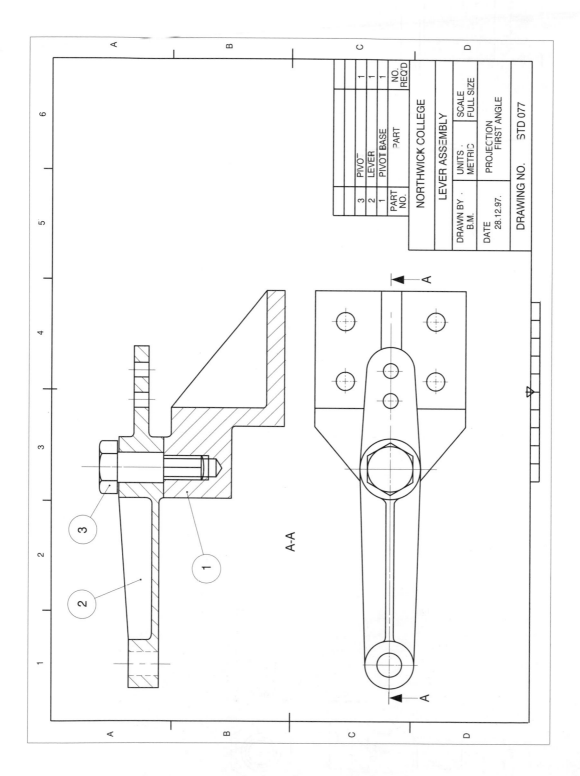

PART NO.	PART	NO. REQ'D
3	PIVOT	1
2	LEVER	1
1	PIVOT BASE	1

NORTHWICK COLLEGE

LEVER ASSEMBLY

DRAWN BY . B.M.	UNITS . METRIC	SCALE FULL SIZE
DATE 28.12.97.	PROJECTION FIRST ANGLE	
DRAWING NO.	STD 077	

A-A

Figure 10.25

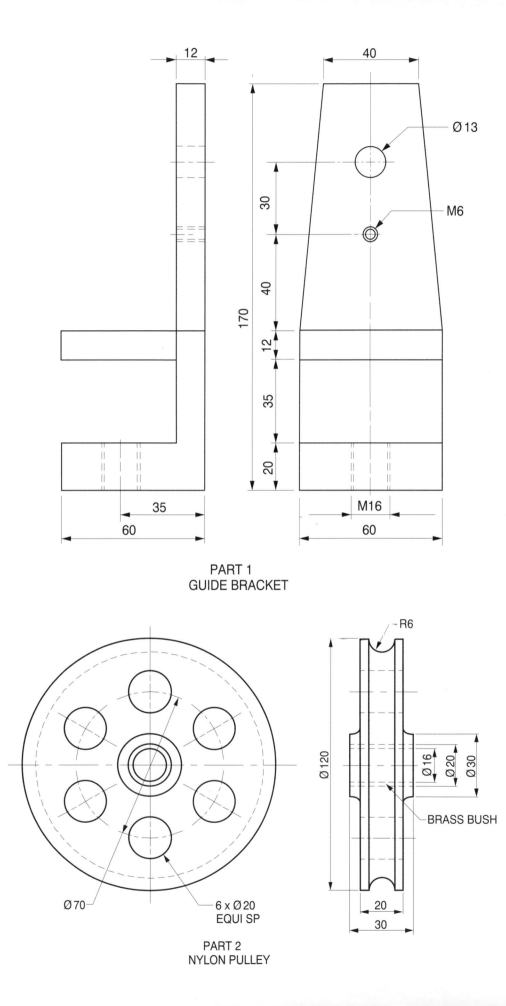

12

40

Ø 13

M6

30

40

170

12

35

20

35

60

M16

60

PART 1
GUIDE BRACKET

R6

Ø 120

Ø 16
Ø 20
Ø 30

BRASS BUSH

Ø 70

6 x Ø 20
EQUI SP

20

30

PART 2
NYLON PULLEY

Figure 10.26

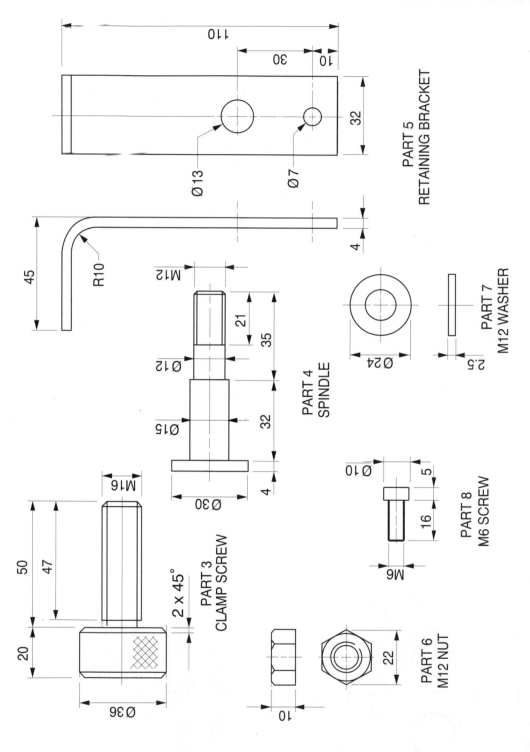

PART 5
RETAINING BRACKET

Ø13

Ø7

R10

PART 4
SPINDLE

M12

Ø12

Ø15

Ø30

PART 7
M12 WASHER

Ø24

M16

PART 3
CLAMP SCREW

2 × 45°

Ø36

Ø10

PART 8
M6 SCREW

M6

PART 6
M12 NUT

Figure 10.26 (continued)

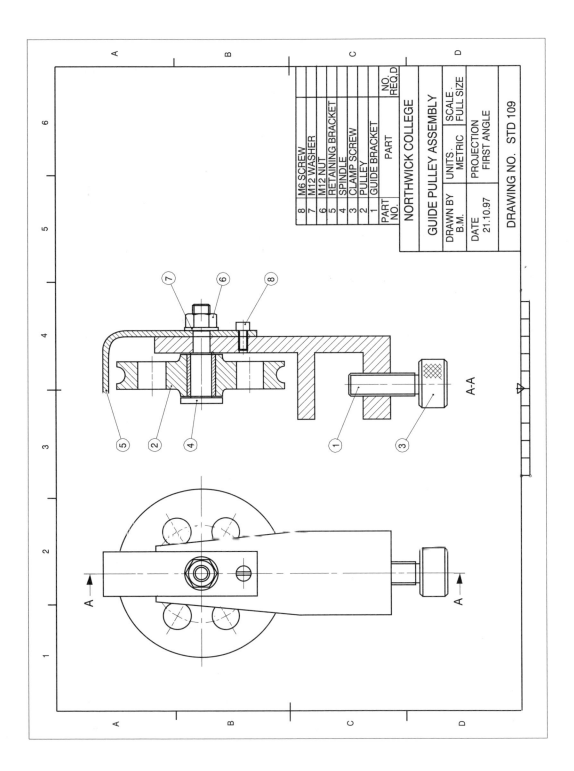

PART NO.	PART	NO. REQ.D
8	M6 SCREW	
7	M12 WASHER	
6	M12 NUT	
5	RETAINING BRACKET	
4	SPINDLE	
3	CLAMP SCREW	
2	PULLEY	
1	GUIDE BRACKET	

NORTHWICK COLLEGE

GUIDE PULLEY ASSEMBLY

UNITS. METRIC	SCALE. FULL SIZE

DRAWN BY B.M.	PROJECTION FIRST ANGLE
DATE 21.10.97	

DRAWING NO. STD 109

A-A

Figure 10.27

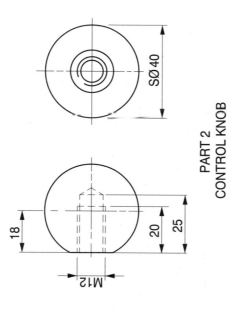

PART 2
CONTROL KNOB

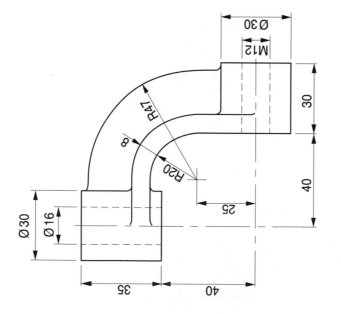

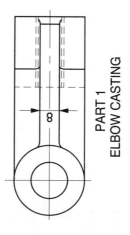

PART 1
ELBOW CASTING

Figure 10.28

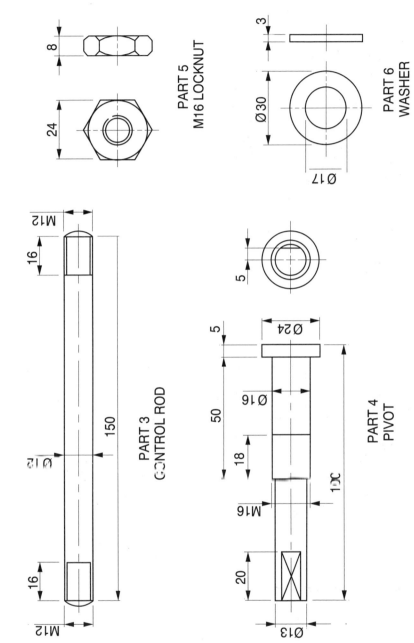

PART 5
M16 LOCKNUT

PART 6
WASHER

PART 3
CONTROL ROD

PART 4
PIVOT

Figure 10.28 (continued)

menu and select **DXF In**. The copied drawing will appear on the screen with the same co-ordinates as before. If the limits of both drawings are identical then the two images will overlap.

To avoid this possibility I have temporarily changed the limits on my framework drawing to 700.00,700.00 and moved the framework vertically upwards outside the range of a landscape A3 drawing. When the command DXF IN is given, the two drawings are positioned on the screen with the framework above the assembly. Obviously the scale of the assembly appears smaller, and you can check before using the MOVE feature that one will fit inside the other. You will appreciate this situation when you observe your first attempt.

Assemble the two drawings and change the drawing limits back to suit an A3 sheet. The result is shown on Fig. 10.17. Clearly we have a large area of unused space. We cannot enlarge the component parts together because the scale enlargement box will overlap the parts list, but they can be done separately and the final solution is shown in Fig. 10.19. A scale factor of 1.3 was used here. Having managed to position the largest drawing in the box, the overall size of the entire assembly of spanner and box can be reduced to fit an A4 drawing sheet.

Exercise 1: cylindrical component clamp

The parts for a component clamp are shown in Fig. 10.20. Draw the separate components, then assemble them together using the COPY and MOVE commands to give the orthographic views in Fig. 10.21.

Exercise 2: trolley wheel assembly

Details of the components are given in Fig. 10.22. Draw the assembly in first or third angle projection with the following views:

- a front view with the wheel bracket positioned as shown on the part 1 details;
- a sectioned end view showing only the details on a vertical plane passing through the wheel bracket and the 10 mm diameter hole for the spindle in part 1;
- a plan view.

Assembly notes

For many components which are instantly recognisable, such as nuts, bolts, washers, screws, rivets, balls, rollers, wheel spokes and spindles, it is customary not to draw them in section. Sometimes spindles do contain internal parts, such as springs and keys, and in those cases a local part section is added for clarity.

In this solution there are several parts and components are cross-hatched at 45°, with the pitch of the cross-hatching reducing in size as the component size reduces. We also try to alternate the angle between adjacent parts if possible. The bracket is fabricated by welding two pieces of material together, so the sectional view indicates that fact. The fillet weld in the front view can be filled in using the SOLID feature in the Draw menu.

Most examination questions require the student to add reference balloons to indicate the part numbers, and these are added by positioning a leader line terminating in a dot from a point within the component. The other end of the leader line extends to a position outside the profile of the assembly and towards the centre of a circle and terminating at the circumference. The part number is placed in the circle; this assembly illustrates this

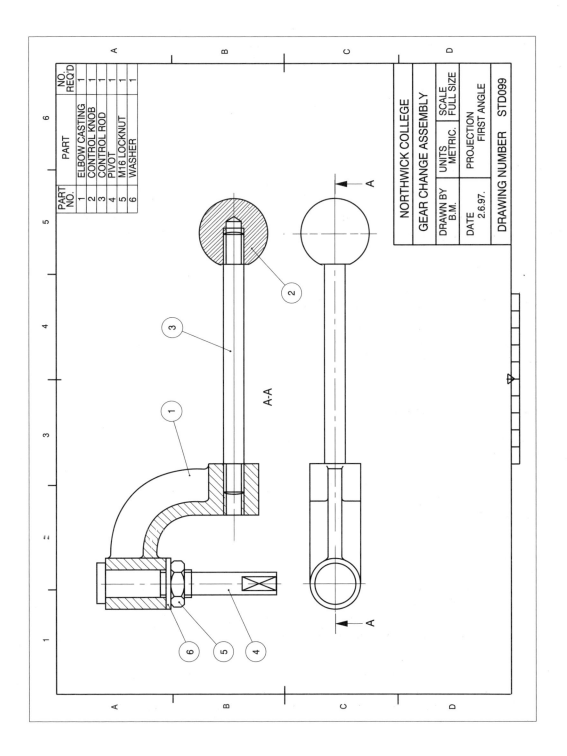

PART NO.	PART	NO. REQ'D
1	ELBOW CASTING	1
2	CONTROL KNOB	1
3	CONTROL ROD	1
4	PIVOT	1
5	M16 LOCKNUT	1
6	WASHER	1

NORTHWICK COLLEGE

GEAR CHANGE ASSEMBLY

DRAWN BY	UNITS	SCALE
B.M.	METRIC.	FULL SIZE

DATE	PROJECTION
2.6.97.	FIRST ANGLE

DRAWING NUMBER   STD099

A-A

Figure 10.29

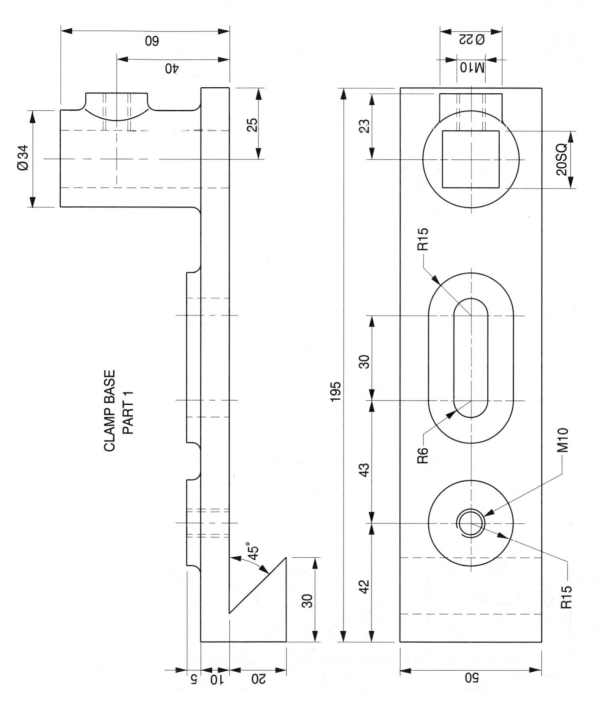

CLAMP BASE
PART 1

Figure 10.30

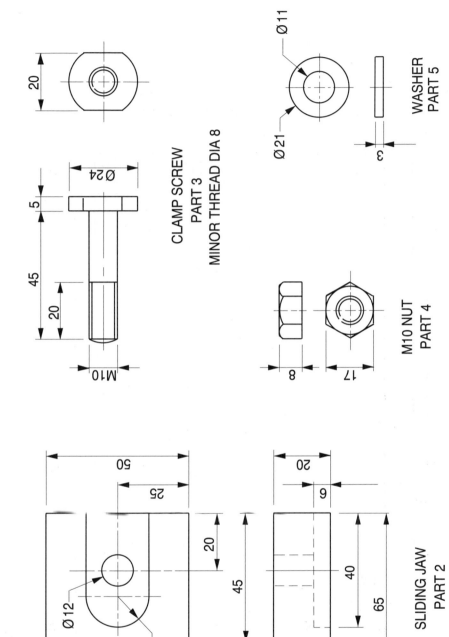

20

Ø24

5

45

20

M10

CLAMP SCREW
PART 3
MINOR THREAD DIA 8

Ø11

Ø21

WASHER
PART 5

3

8

17

M10 NUT
PART 4

50

25

20

Ø12

R11

45

20

6

40

65

SLIDING JAW
PART 2

Figure 10.30 (continued)

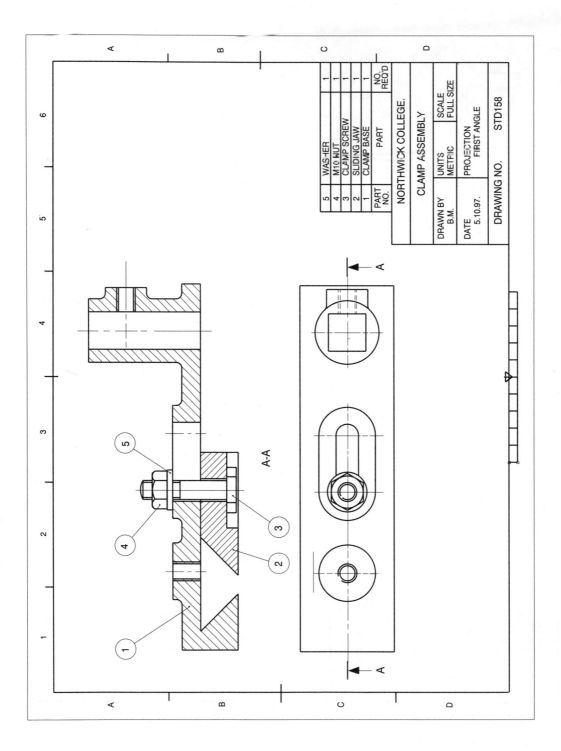

PART NO.	PART	NO. REQ'D
5	WASHER	1
4	M10 NUT	1
3	CLAMP SCREW	1
2	SLIDING JAW	1
1	CLAMP BASE	1

NORTHWICK COLLEGE.

CLAMP ASSEMBLY

DRAWN BY B.M.	UNITS METRIC	SCALE FULL SIZE
DATE 5.10.97.	PROJECTION FIRST ANGLE	
	DRAWING NO.	STD158

A-A

Figure 10.31

addition. Position the part numbers in as neat a way as possible and symmetrically if you can. Figure 10.23 shows the completed first angle projection.

The draughtsman will select sectional views which clearly show the way components are assembled, and as many views as are necessary for clarity.

Exercise 3: lever assembly

Three parts in a lever assembly are detailed in Fig. 10.24. Copy the details and draw a sectional elevation and plan view (Fig. 10.25). Note there are four holes on the pivot base. In order to speed up your draughting, it is only necessary to draw one hole complete with centre lines and then copy the others three. The screw threads are engaged, so cross-hatching does not cover the area between the major and minor diameters. Use a radius of 3 mm for all of the fillet radii. Assemble using COPY and MOVE and remove unwanted linework.

Exercise 4: guide pulley

The dimensioned details in Fig. 10.26 show the component parts of a guide pulley which is clamped to a horizontal surface of thickness 25 mm using the screw (part 3). The pulley

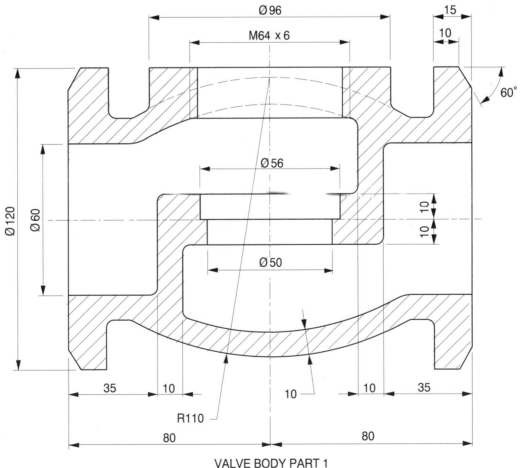

VALVE BODY PART 1
ALL UNSPECIFIED RADII 3mm

Figure 10.32

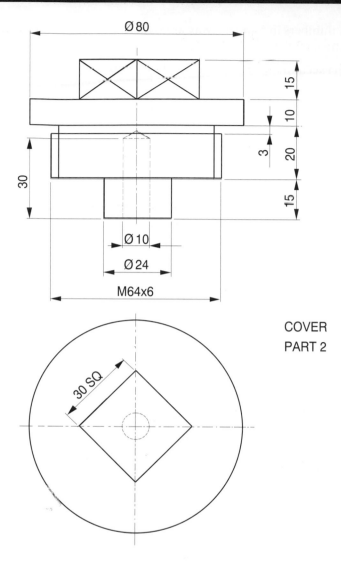

COVER
PART 2

Figure 10.32 (continued)

is used to guide a 10 mm diameter nylon rope. Part 5 keeps the rope in the pulley groove. Draw, using first of third angle projection, the following views:

- a sectional elevation with the parts assembled;
- an end view, which should be an outside view only with no hidden detail included.

A solution in first angle is given in Fig. 10.27. A third angle solution would just involve exchanging the positions of the two views. The object of drawing a sectional view is to illustrate as much detail as possible regarding the construction of a product and how the separate parts are fixed together. If most aspects of the separate parts are displayed, then the use of hidden linework in another view probably does not increase understanding, and its use should be limited. The outside view here provides information regarding width.

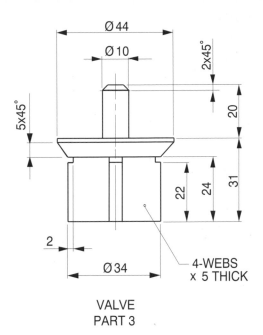

VALVE
PART 3

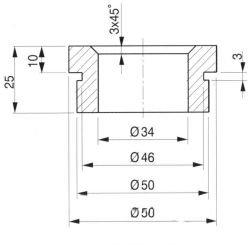

VALVE SEAT
PART 4

Figure 10.32   (continued)

Exercise 5: gear change assembly

Draw the dimensioned details of the parts for a gear change assembly shown in Fig. 10.28. Assemble the components and draw

- a sectional elevation
- a plan view

Note that the details are drawn in first angle projection but if you wish to draw in third angle it is only necessary to reverse the position of the views in parts 1, 2 and 4. The assembly shown, Fig. 10.29 is in first angle projection.

At the end of the pivot the diagonal lines show the standard method of indicating spanner flats, and the width of the rectangle is projected from the end view. Remember

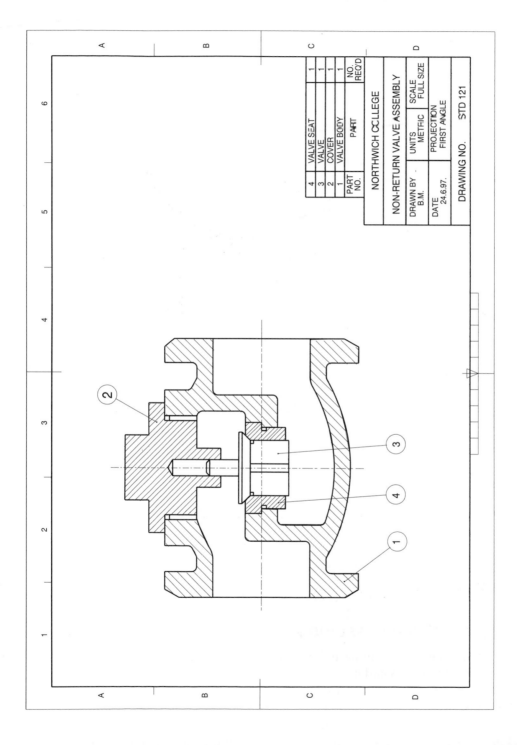

4	VALVE SEAT	1
3	VALVE	1
2	COVER	1
1	VALVE BODY	1
PART NO.	PART	NO. REQ'D

NORTHWICH COLLEGE

NON-RETURN VALVE ASSEMBLY

	UNITS METRIC	SCALE FULL SIZE
DRAWN BY B.M.	PROJECTION FIRST ANGLE	
DATE 24.6.97.		
	DRAWING NO.	STD 121

Figure 10.33

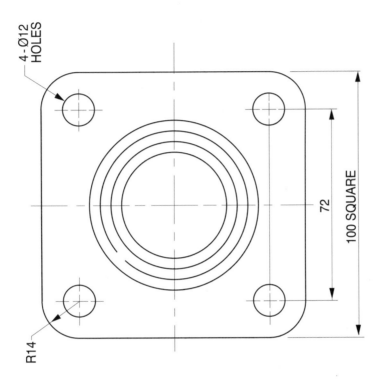

4 - Ø12
HOLES

R14

72

100 SQUARE

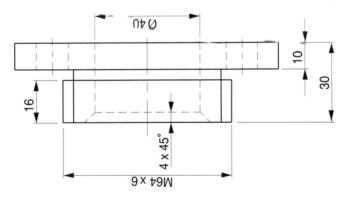

Ø 40

10

30

16

4 x 45°

M64 x 6

VALVE BASE
PART 1

Figure 10.34

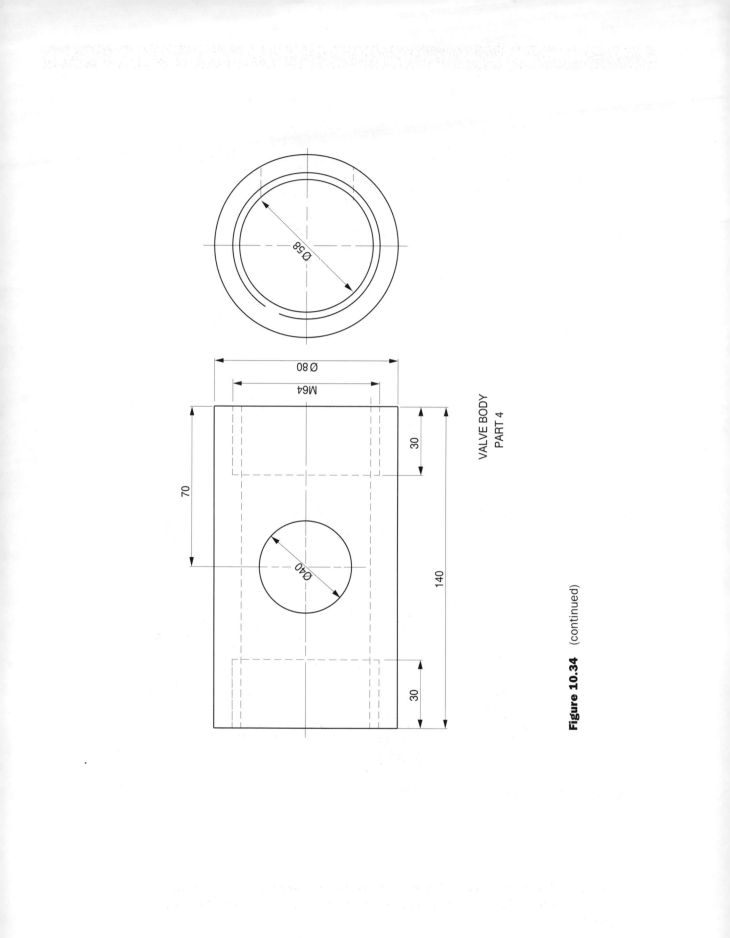

VALVE BODY
PART 4

Figure 10.34 (continued)

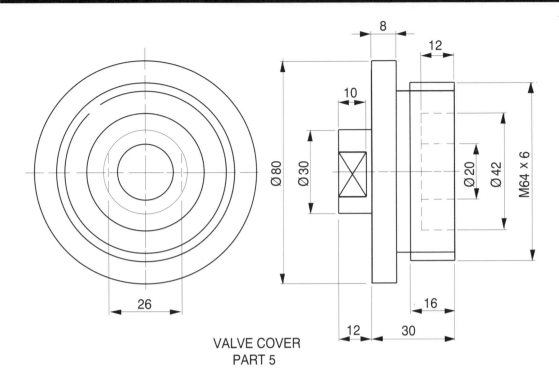

VALVE COVER
PART 5

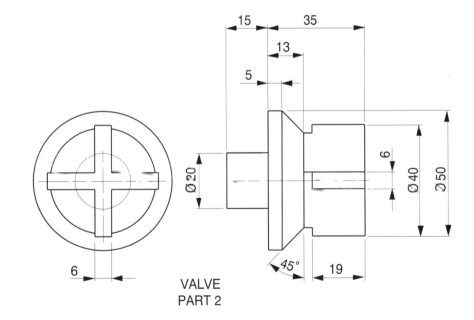

VALVE
PART 2

Figure 10.34 (continued)

not to cross-hatch assembled threads and reduce the width of the cross-hatching on the control knob. The elbow casting also has a thin web which should not be cross-hatched.

For a third angle assembly, simply draw the plan above the sectional view.

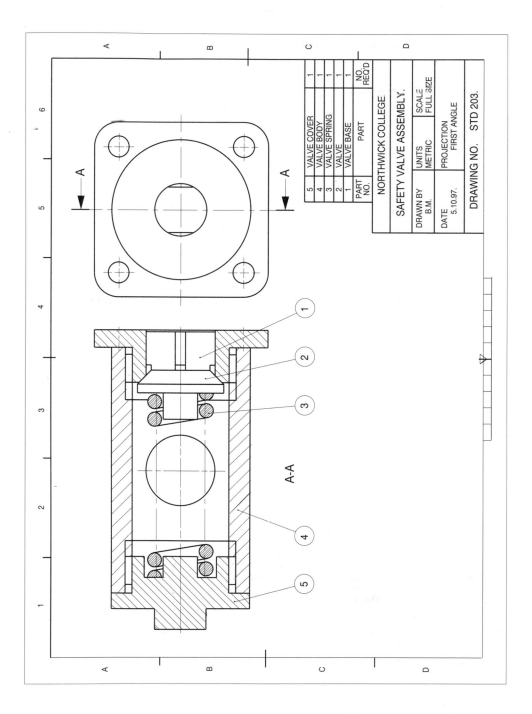

PART NO.	PART	NO. REQ'D
5	VALVE COVER	1
4	VALVE BODY	1
3	VALVE SPRING	1
2	VALVE	1
1	VALVE BASE	1

NORTHWICK COLLEGE

SAFETY VALVE ASSEMBLY.

DRAWN BY B.M.	UNITS METRIC	SCALE FULL SIZE
DATE 5.10.97.	PROJECTION FIRST ANGLE	

DRAWING NO.   STD 203.

A-A

Figure 10.35

Exercise 6: clamp assembly

Copy the dimensioned details of the clamp shown on Fig. 10.30 and then draw the following assembled views:

- a sectional front elevation
- a plan view.

Note that on the clamp base an arc has been drawn where the 22 mm diameter boss connects with the 34 mm diameter post. There is in fact no line here since the fillet radius does not leave a hard line, but the draughtsman would probably add a line for completeness, as shading is not a regular practice on engineering drawings. The arc nearly touches the fillet radii in the front view, and the centre of the arc is projected up from the plan view. Draw an arc passing through these three points in the front view.

Choose **Arc** from the **Draw** menu. Select **3 Point** and the command line reads `Command: _ arc Centre/<Start point>:`. Click at the top near the fillet and command line changes to `Centre/End/<Second point>:`. Click on the centre line and the command line changes to `End Point:`. Complete the arc near to the bottom radii.

Note in the assembly that the bolt, nut and washer are not cross-hatched and the hatching on the sliding jaw has a closer pitch than the clamp base. Hatching for the screw threads is from the minor diameter and covers the major diameter. The solution is drawn in first angle (Fig. 10.31) but a third angle presentation would simply position the plan view above the section plane.

Exercise 7: non-return valve assembly

The component parts of the valve are given in Fig. 10.32. This type of valve would be installed in a pipeline where liquid or gas was required to flow in one direction only. The

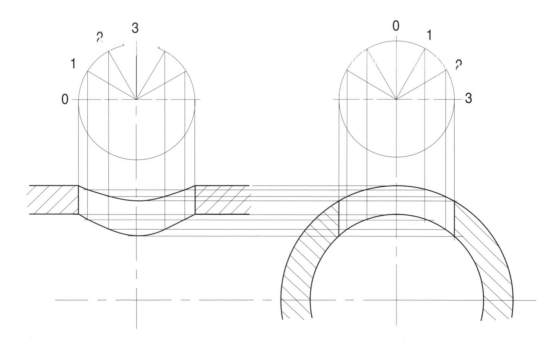

Figure 10.36

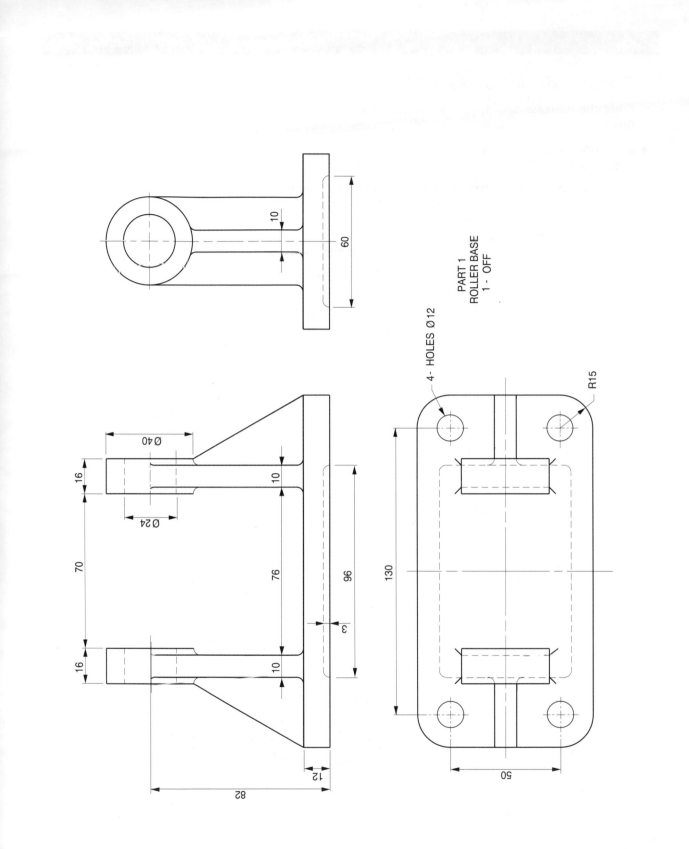

PART 1
ROLLER BASE
1 - OFF

4- HOLES Ø12

R15

Ø40

Ø24

16

16

70

76

96

10

10

3

82

12

50

130

60

10

Figure 10.37

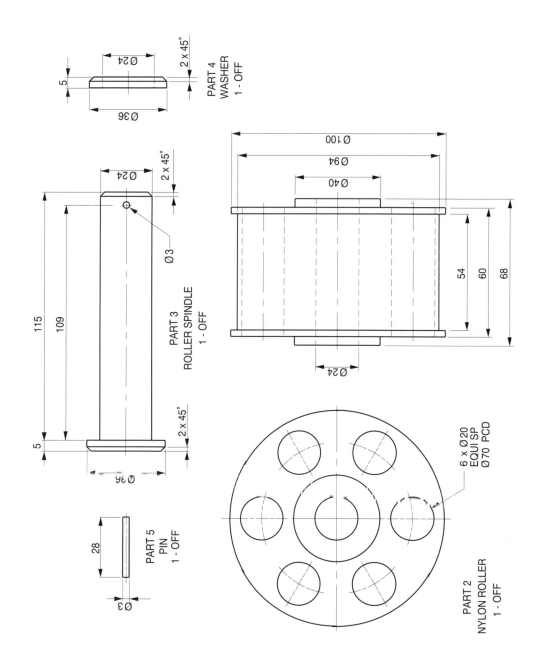

Figure 10.37 (continued)

body is a metal casting and the flanges at each side are designed to be welded into a pipeline. The valve seat (part 4) is a tight fit in the recess at the centre of the body. The valve (part 3) is free to lift up and down in the seat, and is also guided at the top by the spindle which projects into the 10mm bore in the valve cover.

As a draughting and an assembly exercise, draw all of these dimensioned details then save them on to a separate file and assemble the components as previously described. You will see from the solution (Fig. 10.33) that the internal valve would not be drawn in section. Make sure that you use the SNAP feature to move the valve over the centre line of the valve seat. You will need to remove the chamfer lines on the seat. Turn SNAP off temporarily and engage ORTHO while you move the valve vertically down onto the seat.

You will also need to remove two lines in the recess to house the seat in the body and then the previous sub assembly can be lowered into position. Finally screw the valve cover into position and note that when threads are assembled, the cross-hatching is removed in the area where the threads are engaged.

In service, flow will be possible from right to left where pump pressure lifts the valve. Flow would not be possible from left to right since the valve would be pushed firmly against its seat.

Exercise 8: safety valve

The component parts of a safety valve for fitting to a pressure vessel are detailed in Fig. 10.34. If the air in the vessel exceeds the design pressure, the valve is pushed away from the conical seat in the valve base and air is vented through the hole in the body. The valve is held against the seat by a spring, part 3, which is wound from 8mm diameter wire with an inside diameter of 22mm. Draw a sectional view through the valve as shown in Fig. 10.35. The standard for a spring consists of one and a half turns at each end and the centre lines through each side. Springs are normally wound and then ground at each end to give a flat surface for accurate seating. The parts at each end are designed to locate and guide the spring. You can construct the spring from circles and semicircles. Draw one of each and cross-hatch them. Use COPY and MOVE. Alternatively, just construct the spring at one end, then take a copy, use ROTATE and reposition at the other end. Adjust the cross-hatching to suit sizes of components.

Note that if the components had been assembled so that the vent hole in part 4 had been positioned at the top, instead of at the side, and a similar section drawn, then the centre line of the hole would have been positioned on the cutting plane. Figure 10.36 gives the construction for the profile of the sides of the hole, which needs to be plotted, since the sides lie on curved surfaces. The part plans of the holes are divided into six equal segments and plotted as shown. If the points which lie on the hole profiles are lined in by polylines you will get a succession of straight lines between the intersections. However, if you now click on the **Edit Polyline** feature in the **Modify** menu the following options appear at the command line:

```
Close/Join/Width/Editvertex/Fit/Spline/Decurve/Ltypegen/Undo/
eXit<X>:
```

Choose **Fit** and type **F**. The straight lines will be converted to the best curve through the points. This is a very handy feature for drawing graphs, developments and irregular curves.

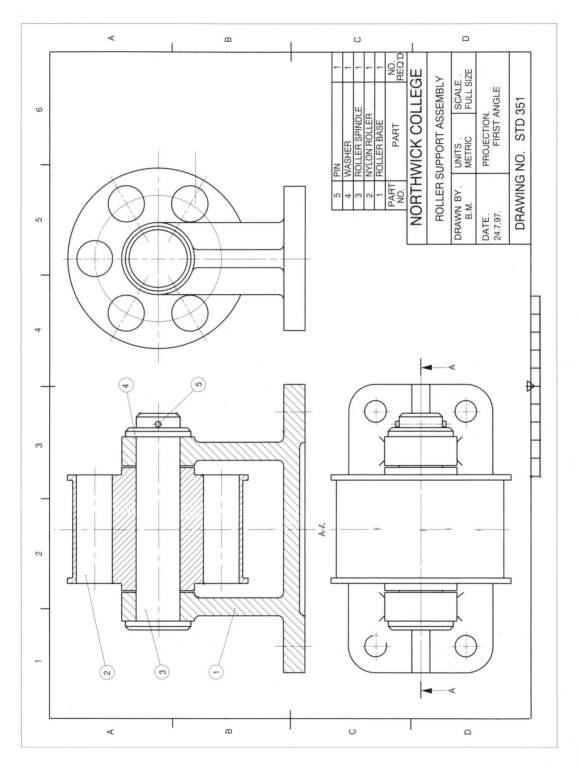

PART NO.	PART	NO. REQ'D
5	PIN	1
4	WASHER	1
3	ROLLER SPINDLE	1
2	NYLON ROLLER	1
1	ROLLER BASE	1

NORTHWICK COLLEGE

ROLLER SUPPORT ASSEMBLY

DRAWN BY. B.M.	UNITS. METRIC	SCALE. FULL SIZE
DATE. 24.7.97.	PROJECTION. FIRST ANGLE	

DRAWING NO.   STD 351

Figure 10.38

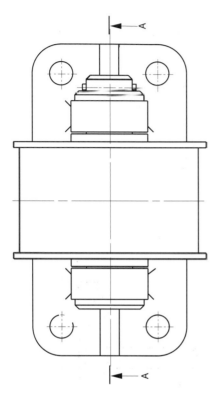

A THIRD ANGLE SOLUTION

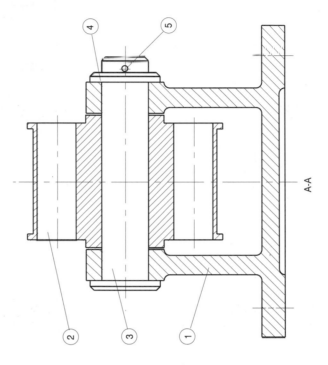

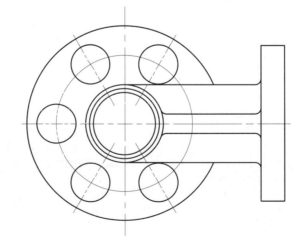

A-A

Figure 10.39

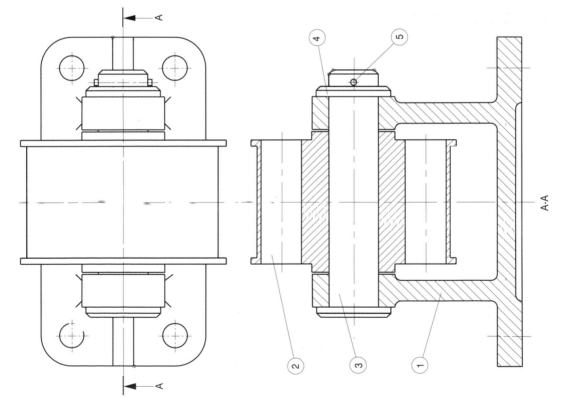

ALTERNATIVE THIRD
ANGLE SOLUTION (SEE NOTE)

A-A

Figure 10.40

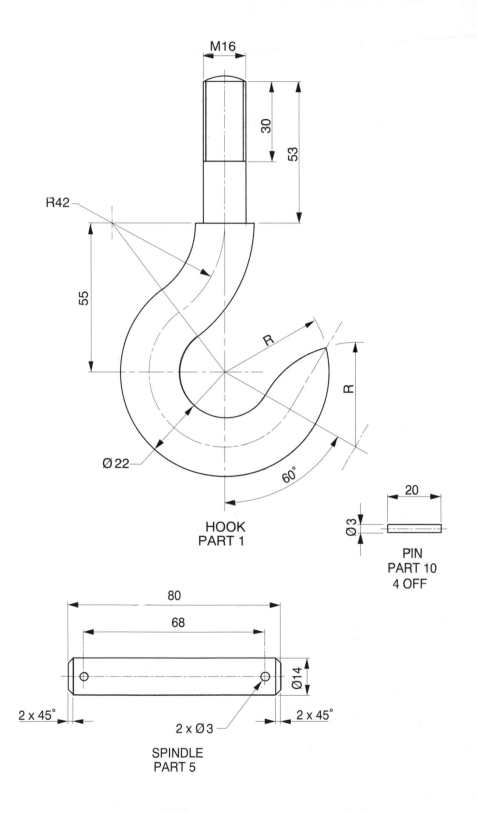

M16

30

53

R42

55

R

R

Ø 22

60°

HOOK
PART 1

20

Ø3

PIN
PART 10
4 OFF

80

68

Ø14

2 x 45°

2 x Ø3

2 x 45°

SPINDLE
PART 5

Figure 10.41

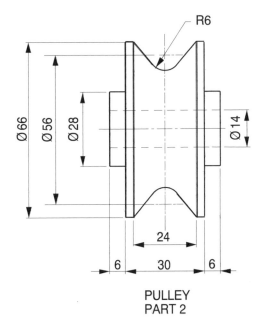

PULLEY
PART 2

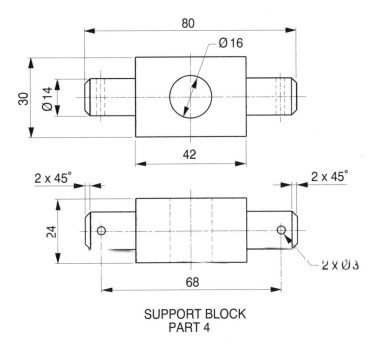

SUPPORT BLOCK
PART 4

Figure 10.41   (continued)

Exercise 9: roller support assembly

The components are shown in Fig. 10.37. We are required to make an assembly drawing consisting of the following views:

- an outside plan view;
- a front view taken as a section on the cutting plane AA;
- an end view.

There are only five parts to this assembly, and in the sectional view we would not

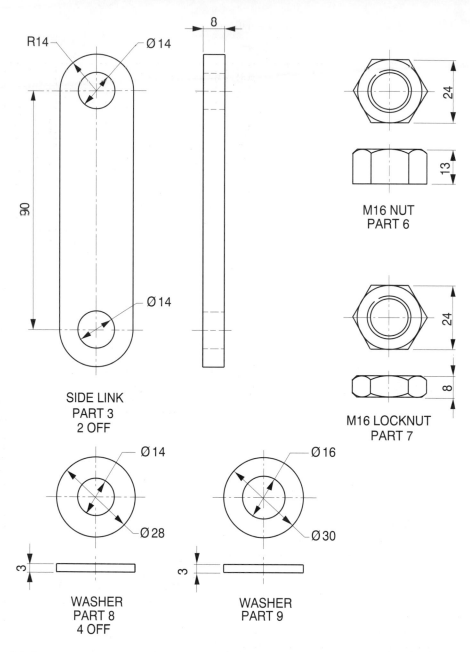

Figure 10.41 (continued)

cross-hatch the spindle. The roller can rotate and it is always the custom to show such parts on a centre line so that the holes appear symmetrically in views. The spindle has a hole for the pin and again this would be oriented in line with the vertical or horizontal centre line. The cross-hatching on the roller, which is a smaller part, would be drawn with a closer pitch than on the bracket and sloping at 45° in the opposite direction.

A first angle solution is shown in Fig. 10.38 and a third angle solution in Fig. 10.39. Note that if an end view is drawn in projection with a sectional view, that the complete view is drawn and not only one half. Balloons have been added enclosing the part numbers. The leader lines to the balloons are directed towards the centres but terminate at the circumference. On the component the leader lines terminate with a small dot.

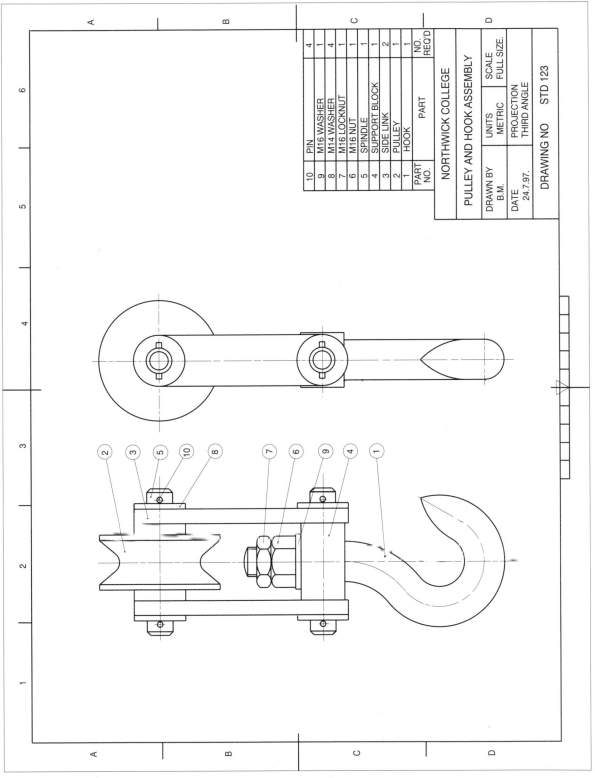

PART NO.	PART	NO. REQ'D
10	PIN	4
9	M16 WASHER	1
8	M14 WASHER	4
7	M16 LOCKNUT	1
6	M16 NUT	1
5	SPINDLE	1
4	SUPPORT BLOCK	1
3	SIDE LINK	2
2	PULLEY	1
1	HOOK	1

NORTHWICK COLLEGE

PULLEY AND HOOK ASSEMBLY

DRAWN BY B.M.	UNITS METRIC	SCALE FULL SIZE.
DATE 24.7.97.	PROJECTION THIRD ANGLE	

DRAWING NO   STD 123

Figure 10.42

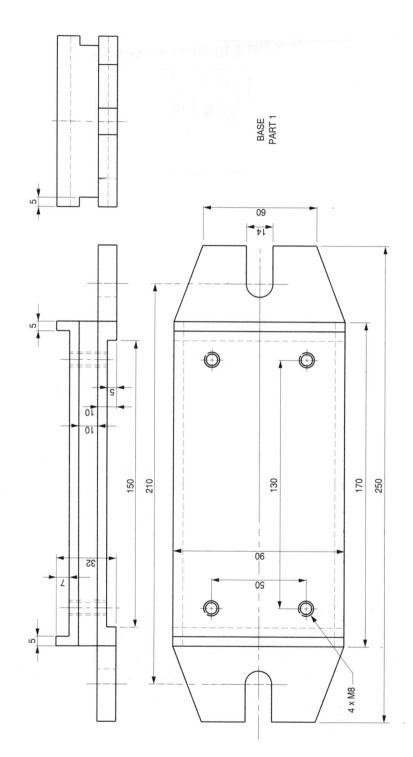

BASE
PART 1

Figure 10.43

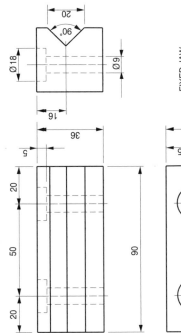

FIXED JAW
PART 4

FIXED BLOCK
PART 2

Ø14
SQ THREAD

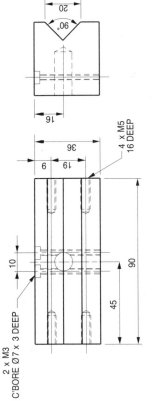

2 x M3
C'BORE Ø7 x 3 DEEP

4 x M5
16 DEEP

SLIDING JAW
PART 3

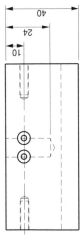

Figure 10.43 (continued)

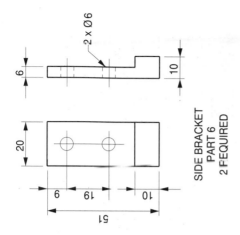

2 x Ø6

SIDE BRACKET
PART 6
2 REQUIRED

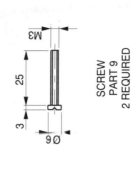

SCREW
PART 9
2 REQUIRED

M3

25

3

Ø6

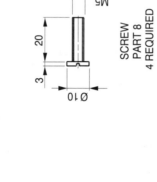

SCREW
PART 8
4 REQUIRED

M5

20

3

Ø10

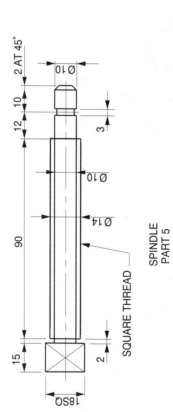

2 AT 45°

Ø10

12 10

3

Ø10

Ø14

90

15

2

18SQ

SPINDLE
PART 5

SQUARE THREAD

SCREW
PART 7
4 REQUIRED

M8

50

5

Ø17

Figure 10.43 (continued)

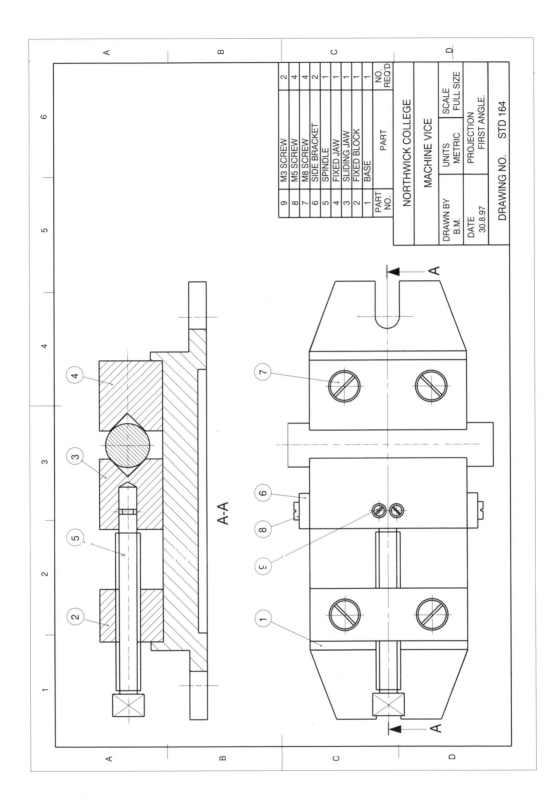

PART NO.	PART	NO. REQ'D
9	M3 SCREW	2
8	M5 SCREW	4
7	M8 SCREW	4
6	SIDE BRACKET	2
5	SPINDLE	1
4	FIXED JAW	1
3	SLIDING JAW	1
2	FIXED BLOCK	1
1	BASE	1

NORTHWICK COLLEGE

MACHINE VICE

DRAWN BY B.M.	UNITS METRIC	SCALE FULL SIZE
DATE 30.8.97	PROJECTION FIRST ANGLE.	

DRAWING NO. STD 164

A-A

Figure 10.44

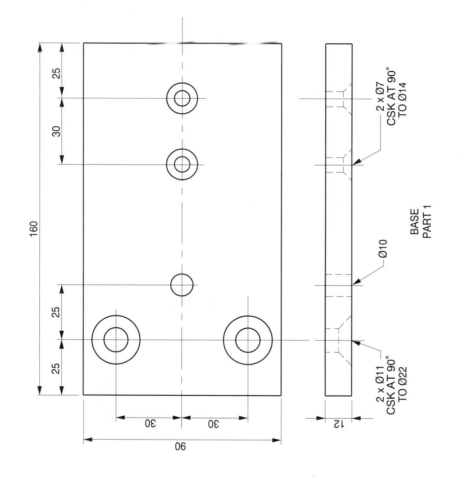

25

30

160

25

25

2 x Ø7
CSK AT 90°
TO Ø14

Ø10

BASE
PART 1

2 x Ø11
CSK AT 90°
TO Ø22

30

30

90

12

Figure 10.45

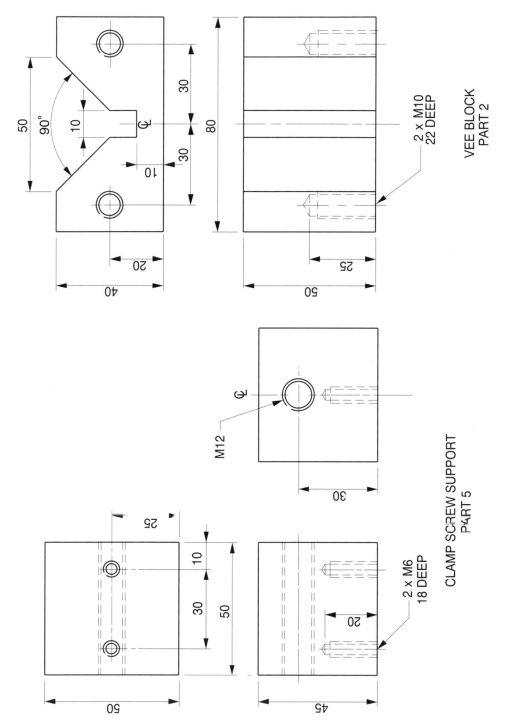

VEE BLOCK
PART 2

CLAMP SCREW SUPPORT
PART 5

Figure 10.45 (continued)

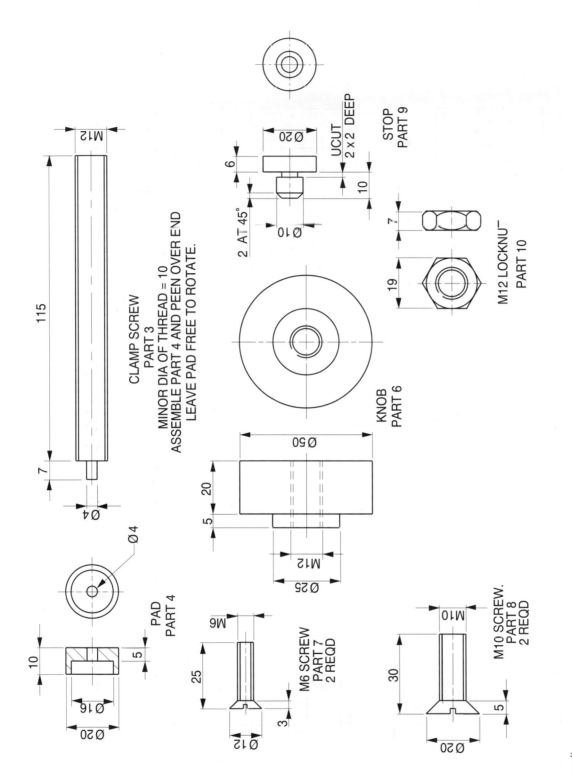

CLAMP SCREW
PART 3
MINOR DIA OF THREAD = 10
ASSEMBLE PART 4 AND PEEN OVER END
LEAVE PAD FREE TO ROTATE.

PAD
PART 4

KNOB
PART 6

STOP
PART 9

M12 LOCKNUT
PART 10

M6 SCREW
PART 7
2 REQD

M10 SCREW.
PART 8
2 REQD

Figure 10.45 (continued)

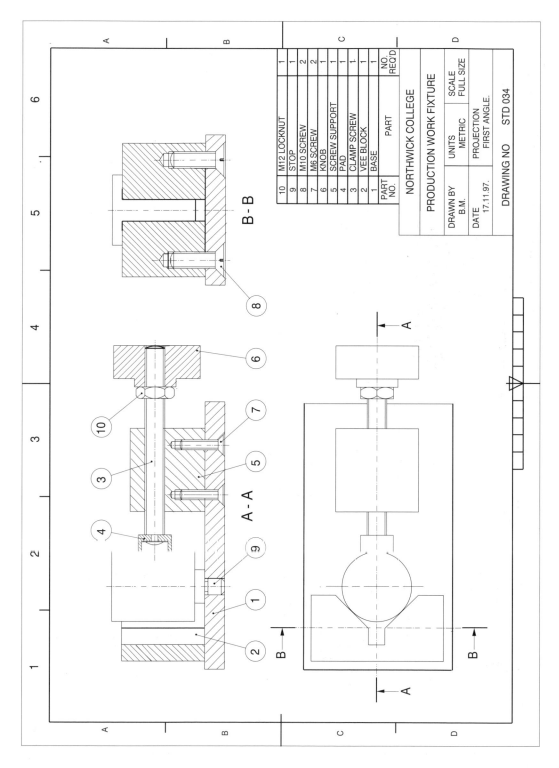

PART NO.	PART	NO. REQ'D
10	M12 LOCKNUT	1
9	STOP	1
8	M10 SCREW	2
7	M6 SCREW	2
6	KNOB	1
5	SCREW SUPPORT	1
4	PAD	1
3	CLAMP SCREW	1
2	VEE BLOCK	1
1	BASE	1

NORTHWICK COLLEGE

PRODUCTION WORK FIXTURE

DRAWN BY B.M.	UNITS METRIC	SCALE FULL SIZE
DATE 17.11.97.	PROJECTION FIRST ANGLE.	

DRAWING NO   STD 034

B - B

A - A

Figure 10.46

You will also see that the alternative arrangement shown in Fig. 10.40 is a valid third angle projection but does not fit into the frame. There is another possible group with an end view to the right of the front view. This is obtained by looking from the right-hand side of the front view, and it conveniently fits into the space. This alternative end view will show the end of the spindle and parts of the pin through the spindle.

Exercise 10: hook and pulley assembly

The components are given in Fig. 10.41 and we are required to draw the details and assemble the parts to give a front and an end view. It is normal practice only to draw sufficient views of components to describe fully their shape and form. Study the details and note that the hook, pulley, spindle and pin do not need another view. The centre part of the support block is rectangular and requires a second view. The second view for the side link and the washer are given to confirm the thickness. All of these items would appear as shown here in first or third angle presentations. The nut is drawn in third angle projection and the position of the chamfer circle in the top view confirms that this is a third angle projection. In the case of the locknut below, this would look exactly the same in first or third angle. The views given do clarify the fact that the locknut is chamfered on both sides, whereas the nut is chamfered on one side only. Note also, for the nuts, that the minor thread diameter is drawn as a complete circle but the major diameter is broken. This is the standard convention for female threads.

The assembly drawing is presented in third angle projection (Fig. 10.42). If, however the tip of the hook was not shown in the right-hand view, then the drawing would be in first angle.

Exercise 11: machine vice

The component parts of a machine vice are given in Fig. 10.43. All components are drawn in first angle projection. Draw the following views of the vice with a circular bar 25 mm diameter and 120 mm long held in the jaws:

- a sectional elevation with the section plane taken along the centre line in the plan view;
- an outside plan view.

Do not include any hidden details in either view.

Note the diagonal lines showing spanner flats at the end of the spindle. The spindle rotates in the sliding jaw and is retained in position by the two M3 screws. The two side brackets guide the sliding jaw along the grooves at each side of the vice. Cross-hatch your solution as shown in the assembly drawing and include the section plane and part numbers. The first angle solution is shown on Fig. 10.44.

Exercise 12: special purpose clamp

Figure 10.45 gives the component parts of a clamp designed to support a cylindrical assembly, which has overall dimensions of 40 mm diameter and 50 mm height resting on a stop (part 9), while further work is undertaken. Draw the given parts and the assembly showing two sectional views and a plan. As previously described, try to use the component views where possible to build up the assembly. The first angle solution is given on Fig. 10.46.

Index

Note: the help button is at the right hand side of the toolbar and provides access to the AutoCAD Help Index. Step-by-step instructions and reference information is available and can be printed if required for later use.